EU ENVIRONMENTAL POLICY

At a time when Europeans across the continent are focused on the EU's future direction, this book provides an important contribution to the current debate. Created for reasons quite unconnected with the environment, the EU has been given a compelling new justification by the success of its environmental policy. A number of factors – including a number of environmental threats that came to prominence in the 1980s, and the new concept of 'sustainable development' – are responsible for pushing EU environmental policy to the forefront of the EU's agenda.

Nigel Haigh, a leading authority on the development and implementation of EU environmental policy, traces the evolution of this policy from obscurity to centrality. Drawing on a range of articles and lectures, he demonstrates how the EU not only has adapted itself to take on entirely new subject matter, but also has contributed to solving problems that individual Member States could not have dealt with on their own. The book also goes on to contextualise the issues throughout the policy's history, and offers insight into the future role of the EU in environmental matters.

This book is a valuable resource for academics and scholars as well as professionals and policy-makers in the areas of environment and sustainability, politics, international relations and European affairs.

Nigel Haigh opened the London office of IEEP in 1980 and was Director until 1998. He has served on the Board of the European Environmental Agency as a nominee of the European Parliament, and on the Board of the Environment Agency (England and Wales).

EU ENVIRONMENTAL POLICY

Its journey to centre stage

Nigel Haigh

To Emma Rose,

In admiration for what you are doing to make a reality of the 'third level' (see marked passage on P. 8.)

Nigel Haigh
September 2022

LONDON AND NEW YORK

from Routledge

First published 2016
by Routledge
2 Park Square, Milton Park, Abingdon, Oxon OX14 4RN

and by Routledge
711 Third Avenue, New York, NY 10017

Routledge is an imprint of the Taylor & Francis Group, an informa business

British Library Cataloguing in Publication Data
A catalogue record for this book is available from the British Library

Library of Congress Cataloging in Publication Data
Haigh, Nigel.
EU environmental policy : its journey to centre stage / Nigel Haigh.
pages cm
1. Environmental policy--European Union countries. I. Title.
GE190.E85H35 2016
363.7'0561094--dc23
2015021687

ISBN: 978-1-138-89030-5 (hbk)
ISBN: 978-1-138-89031-2 (pbk)
ISBN: 978-1-315-71247-5 (ebk)

Typeset in Bembo
by Taylor & Francis Books

In memory of
Konrad von Moltke
1941–2005
Founding Director of the Institute for European Environmental Policy
1976–1984

CONTENTS

ILLUSTRATIONS

Figures

Tables

Box

FOREWORD

What might the passage below, written in May 1982, suggest about the skills and commitments of its author?

> Having persuaded the Chairman of the Lords EEC Scrutiny Committee (Environmental Sub-Committee) that a report on EEC noise policy was timely, I was invited to be the specialist adviser, i.e. to do the work. This involved: reviewing the 23 Noise Directives or proposed Directives, six major EEC policy documents, Pascale Kromarek's comparative study of legislation in France, Germany, Holland and the UK, and various other reports on noise; drumming up written and oral evidence from five Government departments and drafting questions for their Lordships to put to them; visiting the Commission in Brussels with the Sub-committee having drafted a series of questions. My draft 30-page report has now been accepted by the Sub-committee and should be published in July. It may result in a debate in the House of Lords. [*It did: Ed.*]

Plainly, we're faced here with an individual who has the trust of members of an influential parliamentary committee, his judgement commanding respect even in relation to the subjects selected for investigation. It's clear too that he's an expert on both the EU's institutions and developments in environmental policy. He knows his way around Whitehall, Westminster and Brussels – able to manoeuvre within their labyrinthine bureaucracies, and on easy terms with the manners and *modus operandi* of key figures working within them. Most conspicuously of all, he has the skills to orchestrate the exacting forensic work involved in a parliamentary investigation, as well as actually drafting an acceptable final report – all on behalf of a body seen widely as the UK Parliament's most rigorous scrutiny body for European legislation.

At the very least, then, the individual behind the words would seem to be an accomplished, scientifically literate public official, operating behind the scenes in what most of us would regard as the esoteric fields of parliamentary and EU procedures, and European environmental policy. We could be forgiven for wondering whether even a highly experienced parliamentary adviser or high-flying civil servant could encompass the demands of such a role.

But the truth is, the author of this passage, which was originally dashed off as part of a routine internal brief to NGO colleagues at the Institute for European Environmental Policy more than three decades ago, was no Whitehall professional. Rather the contrary – he was an adventurous and innovative *outsider*, a convinced and independent environmentalist, Nigel Haigh, whose work the present volume represents.

As we shall see, such a thirty-year-old snapshot from Nigel's early work offers only a fragment of his story. But what it does do is hint at his behind-the-scenes influence – and, more immediately, the significance of the present book for a fully informed understanding of the evolution of European environmental policy.

To appreciate this better, it's helpful first to step back, to consider the historical background.

Early in the 1970s, several developments of importance for Europe's future had converged. In early 1973, Britain became a full member of the EU, together with Ireland and Denmark – the first expansion of the Community beyond the original six member states. That same year the EU completed its first concrete steps towards an environmental policy, with the creation of a full unit within the Commission with its own programme. Such a coincidence of timing meant that from the earliest stages Britain was an equal partner in the development of the EU's approach towards protection of the environment.

The drive towards new approaches to environmental issues across the western world reflected a shift in the political zeitgeist which had been gathering since the mid-1960s. A wave of scientific concern was interacting with public unease at the scale and significance of pollution and environmental damage across the industrialised world (Minamata disease, the *SS Torrey Canyon*, agricultural pesticides as spotlighted by Rachel Carson).

In the UK, for example, it was popular reaction to late-1960s' pollution scandals that triggered Harold Wilson's out-going Labour government into establishing a standing Royal Commission on Environmental Pollution, as well as initiating the creation of the world's first 'Department of the Environment' (introduced, as things turned out, by the Conservatives in the immediate wake of their 1970 election victory). Out of all this crystallised the early stages of a new *politics* of the environment – one conspicuous manifestation of which was the flourishing environmental NGO 'movement' of the 1970s and 1980s. This involved a growing political sophistication amongst campaign groups and membership organisations, focused on pollution, wildlife and landscape conflicts, as well as wider concerns about energy, ecology and the urgent need for industrial societies to move in directions less profligate and wasteful of resources.

In short, here was a new historical conjuncture with three components – the EU's aspiration to develop an environmental policy; Britain's accession to the Union as a new signatory to the Treaty; and creative stirrings within European civil society in the form of an emergent environmental NGO 'movement' across the member states. It's easy to forget, looking back forty years later, the then novelty of these parallel but interacting developments and the challenges and opportunities they presented to those involved.

Nigel Haigh can be understood as initially a creature of this new combination of circumstances. Following toe-in-the-water involvement in a grassroots environmental campaign, the Anti-Concorde Project, he landed a first full-time berth at the Civic Trust, an independent body dealing with town planning, in 1973. This entailed setting aside an increasingly lucrative conventional future as a patent agent in order to follow his deepest personal convictions. Quickly he became prominent within the scattered *convivium* of independent campaigners and strategists who can now be seen to have constituted the vanguard of the new environmental politics in Britain. Fellow spirits from this period included (both from the fledgling Friends of the Earth) Tom Burke, subsequently an influential policy guru to three successive Secretaries of State for the Environment, and Richard Macrory, a creative driving force behind the dynamic new field of environmental law, now a prominent law professor and QC. A third was the late Marek Mayer, path-breaking editor of the unique environmental intelligence journal, *ENDS*.

But it was Nigel who was the pioneer in relation to Europe. In 1975, the European Environmental Bureau was created, a Community-wide coalition of national environmental NGOs. With Commission support, it provided direct citizen access into Brussels, quickly becoming an indispensable voice. From the outset, Nigel was chosen by the British NGOs as their first representative on its Executive Committee – providing him with a compelling front-line baptism into the complexities of EU policy-making and law.

Subsequently, working closely with the magnetic Konrad von Moltke, there was another first. In 1980, he moved to head up a new London office of von Moltke's Institute for European Environmental Policy (IEEP), with funding from the European Cultural Foundation. Again, this body was a creation of a particular historical moment, colonising a new political-intellectual space as the first public-interest think-tank in the pan-European environmental field. Nigel, Konrad von Moltke and their colleagues created what continues to this day to be a distinctive and influential entity – combining the attributes of independent research institute, international consultancy and environmental reform advocate at European level.

In the 1970s there had been an assumption, not only in the UK, that British environmental regulation was superior to that of continental nations, and so was unlikely to be much influenced by EU membership. Nigel's landmark 1984 book *EEC Environmental Policy and Britain* revealed a rather different story. The UK, like other member states, was being profoundly affected by Europe as new transnational frameworks and principles of regulation evolved, benefiting everything from air quality and bathing beaches to environmental impact assessment and toxic waste control. In the early days Nigel was exceptionally quick to appreciate the full significance of the potential legal regimes that were being created. He saw, in Lord Denning's famous words, that 'the Treaty is like an incoming tide. It flows into the estuaries and up the rivers. It cannot be held back.'

Throughout the 1980s and 1990s, Nigel was thus a considerable back-stage presence in the EU environmental policy sphere, widely respected in key circles for a rare grasp of both day-to-day developments and longer-term trends. As the chapters of this volume show, he and his IEEP colleagues were also active creative contributors to innovations in policy and law, through dialogue with Commission and European Parliament officials, advice to national Parliaments and government departments, and ground-breaking studies taken up by influential politicians. During this period, he also chaired the increasingly influential Green Alliance – and in 2000–05 served two terms as a European Parliament-nominated member of the European Environment Agency.

In the nature of things, activities of these kinds tend to remain obscure to the wider world. But it is important that they be spelt out, as in this volume – not least to assist historians and other analysts to assess the balance of advantage and disadvantage in EU membership, now that such affiliation can no longer be taken for granted as in the past.

To the present reader, at least, the EU's environmental achievements have been strikingly successful in serving the interests of its member states, and continue to be so. Britain in particular has benefited. Whilst it has been a key contributor to EU environment policy, particularly with respect to wildlife and habitat issues, this very success has also acted to prevent its own governments from subsequently undoing such good work, for example in relation to air and water quality. The very existence of EU legislation has secured improvements in health and quality of life in Britain by preventing new governments from chopping and changing policy on issues widely recognised as requiring sustained long-term application.

That this has been the case owes not a little to the remarkable and largely unsung contributions of principled and dedicated individuals like Nigel Haigh, whose forty-year achievement as a truly European citizen merits the appreciation this collection now deserves to bring.

Robin Grove-White
Professor Emeritus of Environment and Society,
Lancaster University
Chairman, Greenpeace UK, 1997–2003

PREFACE

This book is not a history, nor am I a historian. It is based around a number of articles and lectures, nearly all written when I was at the Institute for European Environmental Policy (IEEP) during a period when EU environmental policy was evolving from relative obscurity. Collectively they provide an account of its journey to centre stage by someone who observed it from the point of view neither of a Member State nor of one of the EU institutions. In doing so they may provide material for a future historian. At a time when there is so much discussion about the role of the EU, they show not only how it has adapted itself to take on entirely new subject matter, but also how it has contributed to solving problems that individual Member States could not have dealt with on their own.

The IEEP had been founded in Bonn in 1976 'to inform and guide policy-makers', as its first Director, Konrad von Moltke, used to put it. We accordingly allowed ourselves to have our own ideas and to suggest these to policy-makers, focusing initially on members of the European and national Parliaments, who in those early years had few sources of advice on what was happening in different countries. We saw ourselves neither as academics, nor as a consultancy, nor as a pressure group, though we had attributes of all three. We sometimes lectured at universities; we were paid to undertake research; and by suggesting possibilities, we sometimes helped policy-makers decide what to do. The bulk of the chapters that follow are the result of work whose main purpose was to guide policy. The essential difference between academic research and the products of a policy studies institute is that the latter must be relevant, timely and comprehensible to policy-makers – a phrase as elastic as the word 'policy' itself, and one that embraces more than those who take final decisions.

Each of the articles or lectures reprinted here was accordingly written for a purpose at a particular time. I introduce most chapters with a passage setting the context in which the existing material was written and end with a passage saying what has happened since. In bringing chapters up to date, I have been greatly helped by colleagues at IEEP. David Baldock, the current Director of IEEP, has contributed a final chapter that takes stock of where we are today and looks ahead.

When in 1980 Konrad von Moltke invited me to open IEEP's office in London, he suggested that I write an extended essay on the impact of EU environmental policy in the UK. This led me to study all the EU legislation that then existed and to publish my conclusions. Being part of an institute with offices and partners in several European countries, I learnt how much was gained by having to understand different traditions and solutions to problems. In the following years we continued to study every new item of EU legislation, sometimes participating in the process of its genesis. Colleagues at IEEP have continued this task.

The first chapter confronts the fact that different people see EU environmental policy quite differently, and has been written as an introduction showing how EU policy has matured. Chapter 2 provides a history, and here that word is perhaps the right one, of how the EU acquired the power to enter into agreements on environmental matters with countries outside the EU, and so enabled it to play a leading role in dealing with that most threatening of environmental issues, climate change. That chapter also tackles a fundamental ambiguity about the EU itself that continues to puzzle people and makes it so difficult to understand. Most people will have some idea of how national policy is made in a parliamentary democracy. They also know that foreign policy is something very different. The EU does not fit these two categories. EU policy is neither 'home affairs' nor 'foreign affairs', but has some of the characteristics of both. It is the chapter I am most pleased to have written. Chapter 3, about introducing 'sustainable development' into the EU treaties, is also close to my heart. It was because I had begun corresponding with Konrad von Moltke when we discovered that we had both been calling for the Treaty of Rome to be amended for environmental reasons that I came to join the Institute in 1980.

Subsequent chapters deal with different aspects of EU environmental policy, and in selecting published material I have excluded any that deals primarily with British responses to EU issues as the book is intended for readers throughout the EU, and indeed beyond. Many other countries now look to the EU to see how it tackles environmental problems.

It would be impossible to cover all subjects, but I must acknowledge one major omission. Protecting nature and biodiversity has been an important part of EU policy and deserves full treatment. The few things I have written on the subject do not do it justice and I am not qualified to write a new account.

Nigel Haigh
May 2015

NOTE ON USE OF THE TERM 'EU'

The European Economic Community (EEC) was established by the Treaty of Rome, which was signed in 1957 and came into force in 1958. It was often known as the 'Common Market' after its main task. Its name has changed twice as it has evolved. When the treaty called the Single European Act came into force in 1987 it was renamed the European Community (EC). The name was changed again to the European Union (EU) when the Treaty of Maastricht came into force in 1993. For simplicity, though doubtless to the irritation of purists, I have used the term EU in all the new material in this book, even when discussing events before 1993. In the reprinted older material I have retained the term used at the time of writing.

ACKNOWLEDGEMENTS

My first thanks go to those at the Institute for European Environmental Policy (IEEP), who over many years have built up IEEP's Manual of European Environmental Policy (Farmer 2012), without which it would have been very much more difficult to bring up to date the articles reprinted here. I am grateful to David Baldock, IEEP's current Director, for contributing the final chapter. The book is therefore very much an IEEP project, particularly since most of the previously published material was written when I was at IEEP. When writing the new material, the following at IEEP have provided information, suggestions and insights: Martin Nesbit, Andrew Farmer, Emma Watkins, Sirini Withana and Evelyn Underwood. Pascale Kromarek, whom I first knew when she was at the Bonn office of IEEP in the 1980s, has provided information and made suggestions, as have the following: Adam Hopkins, Frances Irwin, Terry Davies, Debbie Tripley, David Gee, Gwynne Lyons, Ninja Reineke, Michael Warhurst, Julie Hill, David Stanners and Grant Lawrence. Thomas Radice acted as my mentor when I first developed the idea for the book, and also suggested the subtitle, *Its journey to centre stage*. He, Robin Grove-White and Richard Macrory, together with two anonymous reviewers of my proposal, encouraged Routledge-Earthscan to decide to publish. The following at IEEP who understand computers turned the previously published material, together with my drafts, into printable form: Mia Pantzar, Jens Lindblad, Christine Southam and Alejandro Colsa. To all I am grateful. Errors of fact, and very probably errors of judgement, remain mine.

Reference

Farmer, A, ed. (2012) *Manual of European Environmental Policy*. Abingdon: Routledge (available via www.ieep.eu).

PREVIOUSLY PUBLISHED MATERIAL

The publishers and author would like to acknowledge those of the following who first published the material for giving permission to reproduce that material in this book.

Chapter 2 Cooperating with other countries

The paper delivered at Oxford in November 1990 was first published in *International Environmental Affairs* pp 163–180 Vol. 3 No. 3 1991 University Press of New England. It was then published as Chapter 9 in *The International Politics of the Environment* Eds. Hurrell, A and Kingsbury, B 1992 Clarendon Press.

Chapter 3 Sustainable development in the EU treaties

Published as Chapter 3 in *The Transition to Sustainability: The Politics of Agenda 21 in Europe* Eds. O'Riordan, T and Voisey, H 1998 Earthscan.

Chapter 4 Air and acid rain

Published in *International Environmental Affairs* pp 26–37 Vol. 1 No. 1 1989 University Press of New England.

Chapter 5 Water – towards catchment management

The paper delivered at the Financial Times Conference in March 1990 was published in the conference proceedings by Financial Times Conferences. It was then published in *Water Management Europe* 1992 Ed. Garrett, P 1992 Sterling Publications International.

Chapter 7 Chemicals – the Cinderella of environmental policy

The paper delivered in Paris in 2005 was published in the proceedings of the IDDRI workshop *European Proposal for Chemicals Regulation: REACH and Beyond* Ed. Weill, C Paris. Institut du développement et des relations internationales.

Chapter 8 Integrating pollution control

The paper delivered in Berlin in October 1997 was published in German in the proceedings of the conference organised by the German Council on Environmental Law (Die Gesellschaft für Umweltrecht e.V.). It was later published in English in *Environmental Law – Journal of the UK Environmental Law Association* pp 5–9 Vol. 12 No. 2 1998.

Chapter 9 Climate change

Published as Chapter 6 in *Politics of Climate Change – A European Perspective* Eds. O'Riordan, T and Jager, J 1996 Routledge.

Chapter 10 Science and policy

The paper delivered in Amsterdam in March 1998 was published in *Environmental Modeling & Assessment* pp 135–142 Vol. 3 1998 (originally Baltzer Science Publishers, now published by Springer).

Chapter 12 Allocating tasks – subsidiarity

The paper delivered in Brussels in October 1993 and London in February 1994 was published in Environmental Liability CS22 1994 and in *Future European Environmental Policy and Subsidiarity* Ed. Dubrulle, M 1994 European Interuniversity Press.

Chapter 13 The precautionary principle

Box 13.1 was published in *12th Report: Best Practicable Environmental Option Royal Commission on Environmental Pollution* 1988 HMSO.

Chapter 14 Making the legislation work

The paper delivered in Brussels in May 1996 was published in *Community Environmental Law: Making it Work* House of Lords Select Committee on the EC, 2nd Report Session 1997–1998 appendix 5 The Stationery Office.

Except for in Chapter 9, where the reprinted chapter has been substantially shortened, all other previously published material appears as printed or has been very lightly edited to remove obvious errors or ambiguities.

1

SEEING EU ENVIRONMENTAL POLICY

Productive beginnings questioned

The EU had been building an environmental policy for some fifteen years when Jacques Delors began his second term as President of the European Commission in 1989. Long before the EU became involved, several separate strands of what we now call environmental policy had existed in the EU Member States – air and water quality, waste management, nature protection, and land-use planning – subjects that were often handled at a local level with little central government involvement. It was only when the interconnectedness of environmental threats became better understood that some countries began forming specialised ministries in the early 1970s to strengthen and bring these strands together. With the benefit of this experience as well as that of the USA, and inspired by the great UN Conference on the Human Environment of 1972, the EU started to adopt legislation that transmitted to the weaker States some of the policies of the stronger. It was a collaborative process in which EU legislation (mainly Directives, but also Regulations and Decisions) were agreed unanimously by national Ministers who found themselves learning from each other while having to accept compromises. But when it came to dealing with newly emerging topics – reducing acid rain, testing new chemicals, protecting the ozone layer – the legislation the EU introduced was original in all the Member States. Despite the time it took to resolve sharp conflicts over acid rain, it is hard to imagine these three topics being handled as efficiently without the EU's involvement. However productive these beginnings might have seemed, they did not prevent Delors from voicing doubts as he allocated portfolios to his fellow Commissioners. He is said to have told Carlo Ripa di Meana as he asked him to take responsibility for the environment 'I want you to give me an environmental policy. I cannot see an environmental policy. All I can see is a list of Directives.'

Whether Ripa, four years later in any parting interview he may have had with Delors, was able to give an account of how he had risen to the challenge is not recorded, but he could certainly have made an attempt. Delors would have appreciated being reminded that all the Member States had taken a symbolically important step on climate change by agreeing a collective cap on emissions of greenhouse gases – with considerable implications for future energy policies – and that without that step the EU could not have taken the lead in the adoption of a global convention by so many countries, including a reluctant USA. While Ripa was an enthusiastic promoter of action on climate change, he happened to be in post at the very time that policy-makers were confronted with that most threatening of all environmental issues. Delors would certainly have taken satisfaction from European influence being projected globally with the result that people outside Europe began to see that the EU had an important new policy.

Ripa could have gone on to remind Delors how he had raised the visibility of a conflict over polluting motor cars. In order to ensure that small as well as large cars were fitted with catalytic converters, Ripa had taken advantage of the complicated procedures for adopting legislation introduced in 1987 by the treaty known as the Single European Act. That treaty is best known for setting a deadline for the completion of the EU 'internal market' – previously called the 'common market' and now usually referred to as the 'single market'.[1] But it also altered the balance of power between the three EU institutions that together form the EU legislature: the Council (composed of national Ministers); the Commission (which has the power of initiative, so is much more than a civil service); and the elected European Parliament. The Act had not only removed the need for unanimity among national Ministers when the Council adopted legislation relating to the single market (by introducing 'qualified majority voting' which made it impossible for just a few Member States to block a decision), but it had also given the Parliament more power over the outcome. Ripa had persuaded the Commission to drop its previous position in favour of the more stringent emission standards voted for by the Parliament.[2] Since the Council could not agree unanimously to overturn the Parliament's standards, as the new rules allowed it to do, the outcome sent shock waves through the manufacturers of small cars who had complacently assumed that this technology would not be required of them. A new Directive was thus added to the list, but with this difference: manufacturers in an important economic sector found that they could no longer rely on complicit national governments to defend their short-term interests. The EU had very publicly dragged a few reluctant Member States into accepting the higher environmental standards sought by a majority, an outcome that we all – including the car manufacturers – now simply take for granted.

An original Action Programme

While Delors might have been persuaded that new Directives were reaching into important new areas, he could still have questioned whether environmental policy had become more coherent. To answer this more difficult question, Ripa could have explained how the Fifth Environmental Action Programme, which he introduced in 1992, differed from its predecessors.

An Action Programme is the obvious place for presenting an overall view. The first had been called for by the Heads of State and Government in 1972 when they first declared that the EU should have an environmental policy. Subsequent programmes had all taken a similar form: while discussing principles and emerging ideas their main purpose was to outline what new legislation might be brought forward in the years ahead in the separate fields of water, air, waste, nature protection and so on. They were often long and detailed, and accordingly read only by enthusiasts. The lacklustre character of similar programmes in the Netherlands – also divided along bureaucratic demarcation lines – struck Pieter Winsemius when he became Dutch Environment Minister. Drawing on his experience at the management consultancy McKinsey, he suggested something different. He argued that while the earlier programmes might have attracted the attention of specialists, they would never have been read by the Chief Executives or the Boards of companies that generated the problems that the programmes were intended to tackle. He proposed instead that a new Dutch work programme should start by identifying the most environmentally damaging activities – oil refineries, road transport and intensive pig farms, for example – and then say what the Ministry was going to do about them. That, he said, would get their attention.

The Dutch approach had of course to be modified to suit the greater geographical extent of the EU. Instead of identifying discrete problems, the EU's Fifth Action Programme selected five Europe-wide 'target sectors' for attention: industry, transport, agriculture, energy and tourism. The Single European Act had provided the authority for this by stating that 'environmental protection requirements shall be a component of the Community's other policies'. For the first time the Directorates-General of the Commission responsible for these target sectors were on notice that they were likely to face interference. While EU environmental legislation had often caused trouble within the Member States by involving significant costs as well as changes to established rules and procedures, they had made little impact on the other parts of the Commission. These other Directorates-General (DGs) had previously been able to ignore environmental concerns, but were now forced to confront them, as, by extension, were national policy-makers responsible for EU policies in those target sectors too. Anyone with experience of bureaucracies will know quite what that entails. It was an innovation not without risks. If the new Dutch programme woke up Chief Executives, the new EU programme was to wake up other parts of the Commission who might now turn round and bite back.

To make the point that the Fifth Programme was different, it was called 'Towards Sustainability'. This derived from the 1987 report of the UN World Commission on Environment and Development (Brundtland Commission 1987) that had given currency to the concept of 'sustainable development'. The title was intended to convey that EU environmental policy was no longer to be regarded as a self-contained marginal subject; it was now setting its sights on shifting the very core task of the EU. The founding Treaty of Rome had stated this as promoting 'a harmonious development of economic activities', and 'a continuous and balanced expansion' without any suggestion that environmental constraints might limit such expansion. The questioning of the idea that conventional economic growth could be continuous in a finite world of finite resources had seriously begun only in the 1960s and 1970s, and the Brundtland report, reflecting these new ideas, had put forward a form of words that was widely accepted. Development, Brundtland had said, must not be allowed to compromise the ability of future generations to meet their own needs. From then on it became commonplace to distinguish 'sustainable development' from 'economic development' by saying that it was a broader concept supported on three 'pillars' (economic, social and environmental), although this formulation failed to emphasise the tough part of Brundtland's message about future generations. When the Maastricht Treaty entered into force in 1993, the EU's task was accordingly altered to include a reference to sustainable development, although expressed in a rather opaque way for reasons explained in Chapter 3. Clearer wording came with the Amsterdam and Lisbon Treaties of 1997 and 2007, the Amsterdam Treaty stating unambiguously that 'Environmental requirements must be integrated into the definition and implementation of Community policies and activities in particular with a view to promoting sustainable development.'

Delors himself had suggested another way of making EU policy more visible when, in 1989, he called for 'a European system of environmental measurement and verification which could be the precursor of a European environment agency'. Such an Agency began work in 1994 in Copenhagen, more or less at arm's length from the Commission in Brussels, to provide reliable Europe-wide information as a basis for environmental measures.

Strategies proliferate

By 1993, when Ripa ended his term of office, EU environmental policy was looking quite different from when he started, at least to those in EU policy-making circles. Over the next few years it was to continue to change in such a way that Delors might well have said that he could now see a list of strategies, and not just a list of Directives.

When the ink on the Amsterdam Treaty was hardly dry and before it was ratified, the Swedish Prime Minister in 1997 proposed that there be an EU sustainable development strategy, and this was eventually adopted under the Swedish Presidency in 2001 in time for the UN conference on sustainable development in Johannesburg. A Sixth Environmental Action Programme, adopted the next year, was said to provide the environmental 'pillar' of the sustainable development strategy. A series of 'integration strategies' then appeared between 1998 and 2002, each dealing with the subjects of the various specialist Councils – agriculture, energy, transport and so on. This was called the 'Cardiff process' after the Council meeting where it was launched.

The question then arose as to how these various strategies related to the 'overall strategy' for economic and social renewal known as the 'Lisbon Strategy' that was agreed in 2000. This sought to make Europe 'the most competitive and dynamic knowledge-based economy in the world, capable of sustainable economic growth with more and better jobs and greater social cohesion' – wording that conspicuously ignored the environment, depending on the meaning given to 'sustainable growth'. The 2001 sustainable development strategy was said by the Commission to add an environmental dimension to the Lisbon Strategy but as it came later the opportunity was missed to integrate the three 'pillars' – environmental, social and economic – into the first attempt at an all-embracing EU strategy. Partly to overcome this, the Commission began producing a list of 'structural indicators' – over forty at one point – to measure progress with the Lisbon Strategy, and in 2001 seven related to the environment.[3] The total number of indicators was later reduced to a shortlist of fourteen, of which three were environmental: greenhouse gases, energy intensity, and volume of transport. When the Lisbon Strategy was relaunched in 2005, environmental and social considerations were sidelined in favour of a concentration on industrial competitiveness. The Commission then engaged in some wishful thinking when, rather ingeniously, it said that the two strategies (Lisbon and sustainable development) were different but 'mutually reinforcing strategies aimed at the same goal, but producing their results in different time frames'.

The Lisbon Strategy was superseded in 2010 by the 'Europe 2020 Strategy' whose short-term priority is, not surprisingly, to secure an exit from the economic and financial problems that followed the global financial crisis of 2007/08. In the longer term the 2020 Strategy aims to turn the EU into an economy that is *smart* (based on knowledge and innovation); *sustainable* (promoting resource-efficient, greener and more competitive growth); and *inclusive* (high employment, delivering economic, social and territorial cohesion). All these high-level strategies and indicators have had the merit that environmental issues were continuously brought to the attention of the Heads of State and Government at their periodic meetings. Meanwhile new legislation covering the traditional areas of environmental policy, such as those set out in the 'thematic strategies' established by the Sixth Action Programme, have continued, though at a slower pace and often involving refinement of earlier legislation. A subject emphasised in the Seventh Action Programme of 2013 is resource efficiency which is stimulating more discussion of a 'circular economy' (see Chapter 6).

Real world events shift opinion

While there is no doubt that Ripa had done much to move EU environmental policy from self-contained obscurity to being of central importance, several events before his term had begun shifting the views of the public and politicians. Climate change had been placed on the political agenda in 1985 by a world scientific conference prompting the Parliament the next year to call for an EU policy on the subject – the first EU institution to do so (see Chapter 9). When the nuclear power station at Chernobyl in the Ukraine exploded in 1986, sending a cloud of radioactive particles over large parts of Europe, it brought home the message, more immediately than climate change, that pollution knows no frontiers. (The EU promptly introduced safety standards for radioactivity in foodstuffs to prevent differing national standards disrupting trade.) Protection of the ozone layer was another new subject. The hypothesis that it was being depleted by certain synthetic chemicals known as CFCs (chlorofluorocarbons), allowing dangerous levels of ultraviolet radiation to reach the Earth's surface, had been put forward in the 1970s – later to earn its authors a Nobel prize – but in 1987 an international agreement known as the Montreal Protocol required countries to cut production of CFCs as a precautionary measure. No sooner was the Montreal Protocol signed than a scientific consensus developed that CFCs were indeed the cause of the 'hole in the ozone layer', so the ban on CFC production that followed no longer had to be justified as precautionary (see Chapter 13). Suddenly the public became aware that the use of familiar domestic products containing CFCs – hairsprays and refrigerators – had the capacity to threaten the planet. In 1988 the EU finally adopted a Directive to combat acid rain. For many years this had been a concern of Scandinavian countries – which saw themselves as victims of sulphur dioxide blowing mainly from the UK, Germany and Poland. The Directive required the Member States to cut emissions (see Chapter 4). Three of these manmade phenomena – acid rain, depletion of the ozone layer, and climate change – each drew a response that moved EU environmental policy onto a higher plane (see Chapter 11).

A date to mark the transition – 1987

If a symbolic event is needed to mark the beginning of this transition from obscurity to centrality, then the Single European Act that came into force in 1987 best provides it: it formally made environmental protection a component of other EU policies. But no treaty on its own would have led EU environment policy to be taken more seriously by the uncommitted. Nor would an Action Programme, however novel. A necessary ingredient was public concern, and that was strongly reinforced by the four environmental phenomena described above. All four convincingly demonstrated that some problems cannot be handled by individual countries acting alone, and the EU, having been created for reasons quite unconnected with the environment, now found it had a compelling new justification. That the public throughout Europe believed that the EU should have a strong environmental policy partly explains the astonishing success of Green Parties in the 1989 elections for the European Parliament.

Increasing public awareness had also been the objective of the 1987 campaign called 'European Year of the Environment', which was an initiative of Ripa's predecessor, Stanley Clinton Davis. He would never have persuaded his fellow Commissioners to agree to it if they were not sympathetic to the idea that the EU should be paying greater attention to a subject that so concerned the public.

Finally, and perhaps most importantly, 1987 was also the year of the Brundtland report which, as we have seen, gave currency to a worldview-changing concept that influenced subsequent EU treaties (see Chapter 3). For all these reasons, the year 1987 is the best symbolic date for the beginning of the transition period. A date for its end is more difficult to select since a central role for environmental policy always has to be fought for, and is always at risk of being undermined by other pressing short-term concerns, despite a claim that can fairly be made for the EU that it is better at taking a longer view than most national governments. The Amsterdam Treaty of 1997 that paved the way for the sustainable development strategy could be regarded as providing a convenient end marker, but as it took until 2010 before the overarching 'Europe 2020 Strategy' set a long-term goal that more closely embodies 'sustainable development', an end date is best left open. Perhaps we should say that we are still in a period of transition and that backsliding is an ever present possibility (see Chapter 15).

The issues of the ozone layer and acid rain did more than raise awareness. As Chapters 2 and 4 show, they led to EU policy responses that were highly original and provided precedents for approaching the much more intractable problem of climate change (Chapter 9). They gave the EU the confidence it needed in negotiating international agreements and playing a leading role on the world stage. The challenge of developing EU legislation that could work effectively in all Member States had given EU negotiators an important insight into how international agreements could address complex environmental problems.

Three levels of EU policy

The EU policy transition from obscurity to centrality that we have described will not necessarily have been visible to the 'users' or practitioners of EU policy at national or local level: industrialists, public utilities, national and local regulatory authorities, and environmental groups, not to mention the general public. Only politicians closely involved in EU policy, and their advisers and assorted commentators and pressure groups, are likely to have registered it. The looseness with which the term 'policy' is used compounds the very real problem of understanding how different groups see EU environmental policy, so it helps to distinguish three separate levels into which EU policy can be divided:

- the Treaties that give authority ('competence') to pursue certain policies, to legislate in those fields, and to enter into international agreements,
- the strategies and action programmes that set out the intentions of policy, and
- the individual measures, mostly legislation, by which those intentions are to be achieved.

When Delors said in 1989 that he could see a list of Directives but could not see an environment policy, he was effectively saying that he could see the third level but not the second. He would certainly have seen the first level since the Single European Act was the greatest achievement of his long Presidency. The second level was subsequently filled – perhaps confusingly overfilled – by the overlapping strategies mentioned above. It is the third level that most interests the 'users', though they will of course see clearly only the legislation that directly touches them. Most people will regard strategies and programmes as little more than political or bureaucratic talk, however intellectually stimulating they may be and however important as precursors to legislation. While the policy transition from obscurity to centrality can be discerned at the first two levels, it becomes a reality only when it is embodied at the third level, and, most importantly of all, is put into practice by the 'users' in the Member States (see Chapter 14).

Different viewers – different views

Most environmental organisations specialising in, say, nature protection can be expected to know of the Directives protecting birds and habitats, together with only some others. They will take advantage of these to pursue their ends – an example is given in Chapter 2. They may follow EU legislation on climate change, and take an interest in how EU agricultural or fisheries policy is affecting nature, but may well be ignorant of much other environmental legislation. Industrialists need know only of the legislation that directly affects their field. Electricity utilities will know about legislation on acid rain, energy efficiency and climate change; car manufacturers will know of the standards for cars and those affecting manufacturing plants; chemical manufacturers and users will know of the requirements for testing and labelling; retailers will know of the packaging Directive and those setting standards for the products they sell; and so on. Officials responsible for regulating, say, waste management need know only of the waste framework and landfill Directives and a very few more. If they are from a Member State that recently joined the EU, they may be grateful for the availability of a readymade, coherent legislative package relevant to their field. If they are from a Member State with a strong existing tradition, they may be irritated if new EU legislation forces them to do things slightly differently. These 'users' are all bound to see EU policy differently.

Some 'users' pay attention when new legislation is being developed. When a 'thematic strategy' on air pollution, that was to lead to new legislation, was being discussed between 2001 and 2005 over 100 stakeholder meetings were held and over 11,500 responses were received to a web-based consultation. Many other Directives or Regulations have been the subject of intense lobbying of the Parliament, the Commission and national governments. Those likely to be affected by new EU legislation now have their eyes wide open.

Officials responsible for developing EU legislation will also have their own way of looking at it. When the Commission formally proposes a Directive, it is studied in a Council working group of officials from the Member States before it goes to a formal Council meeting. The Commission, in making its proposal, will be acting in what it sees as the European interest, and its ideas are likely to have been prefigured in an Action Programme or strategy or developed in response to calls by the Council or Parliament. Officials in the working group should also be acting in the European interest but will see their primary responsibility as assessing how it will affect their national interests and how it fits their national traditions and practices. Officials from different countries are therefore likely to see the proposal slightly differently and will suggest amendments accordingly. Once the working group has finished its deliberations and agreed a text, or has identified unresolved points or key issues for political choices, the proposal will go to the formal Council which will have to find any compromises that may be necessary. Meanwhile the Parliament will have used its powers to propose amendments that may well reflect the views it will have received from the public and interest groups. Many of those involved in adopting the legislation tend to be so concentrated on the wording of the text before them, and the need to come out on the winning side of a short-term debate on direction, that they sometimes lose sight of how it will come to be implemented: that will be someone else's job later on (see Chapter 14).

EU environmental policy accordingly lies very much in the eye of the beholder. The list of beholders so far mentioned above is incomplete and does not include countries outside the EU, which often look to EU legislation as a model; nor national Parliaments, which now have a formal role in commenting on 'subsidiarity' (see Chapter 12). One other participant must be mentioned. The European Environment Agency (EEA) collects data from the Member States on the quality of the environment, and takes numerical standards in EU legislation as a reference. It will therefore pay particular attention to Directives that contain such numerical standards to the neglect of those that lay down procedures whose effectiveness can be assessed only qualitatively. Even a body charged with gathering Europe-wide information to protect the environment will have its own view of EU policy and will not necessarily see it whole.

Finally there is the general public. Public opinion is a term that is imprecise enough when used within one country but is even more imprecise when used within the EU. Some argue that there is no such thing as a European public opinion, given that opinion is grounded in national traditions, language and history and is informed by national media and political processes. Since, for cultural or geographical reasons, different priorities are often given to different aspects of the environment, it is no surprise to find that different items of EU legislation can be viewed quite differently in different Member States. The point was well made by the southern European who reluctantly admitted that many of his fellow countrymen thought it acceptable to eat protected wild birds, unlike outraged northerners who, as he said, poisoned them with pesticides instead. Despite such differing perspectives, the regular social surveys conducted by Eurobarometer show that the public in Europe generally understands the reasons for EU environmental policy and views it positively.

As this book presents a personal view, let me end this chapter with a personal anecdote. I was once visited by a senior official from a country that was in the process of joining the EU. He was travelling around Europe talking to his opposite numbers in Environment Ministries to learn what to expect. After his country had joined the EU we met again and he thanked me for making a point that he had found helpful and that no-one else had put to him. I had apparently told him that what distinguished the EU from any other international organisation was that it had the power to legislate and that EU environmental policy was embodied in that legislation. As this is a simple statement of fact, I was left wondering how it was that his experienced opposite numbers had failed to draw his attention to this absolutely key feature of the EU. Did they regard it as self-evident and so not worth mentioning? Or were they so involved in the current policy debates that they had lost sight of the list of Directives that Delors had once said was all he could see? As following chapters will show, all the advances in EU environmental policy have been embodied in legislation. Since the most important subject of all – climate change – involves operating on the world stage, we start with how the EU acquired the powers to do so.

Notes

1 It is widely believed that the Single European Act is so named because it established a 'single market'. But the Act uses the term 'internal market' and not 'single market'. The 'Act' – an unusual name for a treaty – acquired its name for the prosaic reason that it combined two texts that had different origins, one amending the Treaty of Rome (Title II) and one dealing with cooperation in the field of foreign policy (Title III).
2 See Chapter 10 for a fuller description.
3 They were: greenhouse gas emissions; the share of renewable energy; volume of transport per unit of GDP; the split between transport modes; exposure of urban populations to air pollution; the volume of municipal waste collected; and energy consumption per unit of GDP. It will surprise many that nature protection was not included. Although halting the decline of biodiversity by 2010 had been set as a target by 2001, the difficulty of measuring progress meant that no indicator could then be produced.

Reference

Brundtland Commission (1987) *Our Common Future: Report of the World Commission on Environment and Development*, Oxford: Oxford University Press.

2

COOPERATING WITH OTHER COUNTRIES

When discussing the EU's entry into 'foreign affairs', it is as well to be as clear as possible about the differences between 'home affairs', 'foreign affairs' and 'EU affairs'. Readers who are confident they know the differences will wish to skip over the next few paragraphs and go straight to the paper below. It was first presented as a lecture in a series dealing with the international politics of the environment (Haigh 1992).

In a parliamentary democracy of the kind found in all EU Member States, people are likely to have a fairly clear idea of the basic elements that enable their country to function. They will know the name of their Head of State, be he/she an elected President possessed of wide powers, or a constitutional Monarch or President with limited powers. They know that there is an elected parliament that makes laws that guide the conduct of individuals and companies; that political parties offering different policies compete to form a government; and that the government administers those policies, raises taxes, and introduces new laws. They know there are Courts that uphold the law and resolve the conflicts that arise between individuals and legal entities as they go about their business. They also know there are local and other authorities that provide services and regulate certain activities. We refer to all these matters as 'home affairs' or 'internal affairs' or 'home policy', and we are so familiar with them that we take them for granted. Those states that have regions with legislative powers, whether or not the country is formally called a federation, are used to the existence of more than one legislature with powers shared between them. People in such states know that it is the national government that is responsible for foreign affairs.

Everyone knows that nation states have relations with other states, some friendly, others less so. This involves Ministers and officials talking with their opposite numbers in other states. If 'home affairs' is about the relations between a nation state and the individuals and other entities within it, 'foreign affairs' is about the relations between sovereign nation states. Traditionally the main subject matter of international relations has been defence and trade, but increasingly the environment is discussed. When these states wish to give a binding character to what they have agreed, they will enter into a treaty or convention.[1] After a treaty is negotiated – sometimes in secret following the traditions of diplomacy – it is signed but then has to be ratified according to the constitutional arrangements of the contracting states before it enters into force. National Parliaments will not be involved in the negotiations but may have to approve ratification, and it is not unknown for ratification to be refused. Once a treaty is fully ratified it binds the parties and becomes part of international law. In states called 'monist' the national courts recognise international law as if it were national law. Other states do not, and ratification may have to be delayed until their national legislation is amended to fulfil the obligations in the treaty. However, if a state fails to fulfil the obligations it has entered into, there is usually very little the other contracting states can do. We refer to all these activities as 'foreign affairs' or 'external affairs' or 'foreign policy'. It is fair to say that even the well educated will usually have only a vague understanding of the processes involved, perhaps because they see them as remote from their lives.

A growing aspect of foreign affairs has been the creation of international organisations which individual states can apply to join. Some are global, such as the United Nations (UN) or the World Health Organization (WHO). Some have a specialised function, and some are limited to certain states, such as the Organisation for Economic Co-operation and Development (OECD). Most international conventions on the environment will create a specialised Commission to administer the convention and convene a periodic 'conference of the parties' (COP). Of particular relevance for environmental policy in Europe have been the United Nations Economic Commission for Europe (UNECE), the OECD and the Council of Europe.[2] These organisations can make pronouncements, but if these are to be binding their member countries must usually enter a treaty or convention as described above.

When it comes to the EU, most people are even more unsure of how it works than they are about foreign policy. To regard the EU as another international organisation is to misunderstand it, and to regard EU policy-making as an aspect of foreign affairs leads to confusion. I experienced this confusion when I was asked to give the following paper as one of a series at Oxford University on the international politics of the environment (Hurrell and Kingsbury 1992). I assumed, from the title given me, that I was to speak on the EU's relations with countries outside the EU – its external policy. But the organisers wanted me also to cover the making of EU legislation that binds only its Member States – its internal policy. In doing so, I had to answer the question whether the EU's internal policy should also be regarded as foreign policy. The answer given was that 'Although EC policy has the attributes of foreign policy during its formulation, it then becomes integrated with home policy in its implementation.' There is no denying that the EU is not easily understood, but it is best to regard EU policy-making as neither 'foreign' nor 'home' policy-making, but as something unique. There are thus three categories of policy-making: home affairs, foreign affairs and EU affairs, each following its own rules.

The paper briefly describes how the EU functions internally before discussing how it acquired the power to become a party to international environmental conventions and so to play a significant role in the protection of the ozone layer. This was to pave the way for it to do the same for climate change. The paper was given in 1990 when the EU was still called the EC and when the Parliament had fewer powers than today. The general argument is not affected.

The European Community and international environmental policy – paper delivered at Oxford, November 1990

> Within their respective spheres of competence, the Community and the Member States shall cooperate with third countries and with the relevant international organisations.
>
> *Article 130R (5) of the Title of the Treaty of Rome concerning environmental policy*

The ambiguous character of the European Community

One of the marks of nation statehood is the ability to enter into agreements with other nation states. This ability is shared by the European Community (EC) even though it is not itself a nation state but is a 'Community' established originally between six, and now twelve, nations each of which continues to express its own sovereignty by pursuing its own foreign relations. It is therefore hardly surprising that when the EC and its Member States negotiate with other countries there is some ambiguity about their relative roles.

The tasks, powers and institutions of the EC are set out with a fair degree of clarity in the Treaty of Rome which created the EC in 1957. By establishing a common market the founding fathers sought to achieve closer relations between the Member States. Although there are many who would like to see the EC evolving into a kind of United States of Europe, it is easy to demonstrate that the EC has not yet achieved the quality of nation statehood. For a start it has no head of state, nor a constitution adapted for dealing with all eventualities, nor even the power to raise taxes directly from citizens. All these are characteristics normally regarded as essentials of nation statehood. On the other hand, the EC is more than an international organisation established between nations to pursue some prescribed activity without fundamentally ceding any of their sovereign powers. What distinguishes the EC is its possession of institutions able to adopt legislation which directly binds the Member States without further review or ratification by national institutions. In the environmental field the extent of the legislation is such that it is now impossible to understand the policies of any EC Member State without understanding EC policy. Member States are no longer entirely free to pursue their own policies, whether at home or abroad.

The EC legislature is composed of the Commission acting together with the Council (composed of national Ministers). The Commission proposes the legislation and the Council adopts it after receiving an opinion from a Parliament directly elected by the citizens of the Member States. The European Court of Justice has the power to find against a Member State that fails to apply EC legislation correctly and the Commission does not hesitate to bring cases before the Court. The powers of the EC to legislate are certainly confined to subjects prescribed by the Treaty of Rome, but a generous interpretation has often been given to the Treaty, and the EC was able to adopt an environmental policy in 1972 despite the absence of express powers. These were not introduced until the Treaty was amended in 1987 by the Single European Act.

The influence of the EC on its Member States has increased rapidly at some times, less so at others, and in certain fields there has sometimes been near immobility. In December 1990 an intergovernmental conference began work on a revision of the Treaty for the purposes of monetary union and political union, which could substantially extend the EC's powers. One of the tasks of the conference is to give greater precision to these concepts. As the powers of the EC increase, and its internal policies more deeply influence the Member States, so the EC is strengthened in its ability to act in its own right on the international stage. This will apply in the environmental field as it will in others, but since the result is a diminution in the ability of the Member States to act on their own, the path is unlikely to be smooth. This evolution of the EC coincides with the emergence of the environment as a major new subject of international affairs. The EC has already played an important role on the world stage in the protection of the ozone layer and is beginning to do so with global warming. But as the policy responses to global warming may have profound effects on many aspects of national life, the exact role of the EC will not be decided easily.

Is European Community policy foreign policy?

If the role of the EC relative to its Member States in external affairs is one ambiguity, another one has also to be confronted. Is EC internal environmental policy to be regarded as international on the grounds that it involves relations between several nation states, or does the title of this paper confine discussion to the EC's external relations?

To regard internal EC policy as within the scope of this paper as the organisers have wanted is a perfectly tenable view, even if by accepting it one compounds a confusion that commonly surrounds the EC.

It is as well to get to the bottom of this confusion since it is particularly important in the environmental area, where responsibility for implementing policy – and thereby helping to make it – frequently rests with local authorities or other subnational bodies who would not normally expect to be involved in foreign affairs, but who increasingly find themselves having to come to terms with EC affairs. One of the reasons why the EC is so difficult to understand is that it does not fit the simple model of public policy under which it is commonly divided into home and foreign affairs. EC policy-making shares so many of the characteristics of foreign policy-making that it is easy to think of it as such. For a start, it involves other countries and is usually made abroad. Despite the growing role of the European Parliament, EC policy is prepared largely in secret following the traditions of diplomacy, and legislation is adopted by a Council of Ministers behind closed doors. The process is thus much more like treaty-making than the open process of adopting national legislation which is the hallmark of parliamentary democracies. It is still possible for EC legislation to differ in significant respects from the proposal originally published by the Commission without outsiders knowing who was responsible for the changes or why.[3] But EC legislation once adopted is quite unlike a Treaty and can have the same force as national legislation. One form of EC legislation – the Regulation – is directly applicable by national courts just as if it were national legislation. Another form – the Directive – binds national governments as to the ends to be achieved and can also be applied by national courts in some circumstances.[4] Both these forms affect internal affairs without further review or ratification by national Parliaments. If a Member State fails to fulfil the obligation set out in an EC Directive, the Commission can bring an action before the Court and so draw attention to the failure. All Member States effectively now have two legislatures, and the higher legislature (EC) can influence internal decisions just as does the lower (national) legislature.

The implications of this point are best illustrated with an example. Duich Moss is a peat bog on Islay, one of the islands off Scotland. Islay is the seat of a distillery producing a single malt whisky that is renowned for a flavour imparted to the malted barley by the burning of peat. A few years ago the distillery wished to expand its production and applied for planning permission involving a new site for digging peat, and an access road. The planning application was opposed by the Nature Conservancy Council (NCC), the UK Government's official adviser on nature conservation matters, and by a private body, the Royal Society for the Protection of Birds (RSPB). In winter 4 to 5 per cent of the entire world population of the Greenland White Fronted Goose is to be found feeding at Duich Moss.

This is a typical case of a conflict arising between two interest groups, the resolution of which it is one purpose of home policy to provide for. In this case, national rules exist and are administered locally, unless exceptional circumstances suggest that the central government should intervene. Under the town and country planning laws, permission to develop land is granted or withheld by the local authority, after certain matters have been taken into account, and after interested parties have had a chance to make representations. In this case, the authority decided in favour of the distillery. In the ordinary course of events that would have been the end of the matter, but the RSPB, having lost the battle under the rules of home policy, proceeded to play an EC card. They complained to the EC Commission in Brussels that the United Kingdom was in breach of the EC Directive on the conservation of wild birds, which requires Member States *inter alia* to classify special protection areas for the conservation of named species of bird. The RSPB argued that under any reasonable criteria, Duich Moss should have been classified as a 'special protection area' and that had it been so classified the outcome of the planning application would have been different.

Complaints to the Commission from individuals, local authorities, and interest groups, are a growing feature of the implementation of EC legislation (see Chapter 14). The Commission has been encouraging this as one way of keeping itself informed. The Commission registers these complaints and, since it has no inspectorate of its own, it usually writes to the Member State asking for an explanation. In this case, however, the Commission official decided to go and see for himself. As a matter of courtesy, the British Foreign Office was informed of his intended visit and was thereby thrown into a state of mild panic. Nothing like this had ever happened before in the EC. Did the official have the right to come? Was his visit desirable? Could he be stopped if he insisted on coming? One can imagine the questions being asked. There is no doubt that the official could visit Islay as a holiday maker to watch birds, but he was intending to come as a representative of the EC institution that is the guardian of the Treaty, in order to investigate a complaint, by a private body, against a Member State for failure to fulfil an obligation under the Treaty. The Treaty gives no guidance whatever on the right to visit.[5]

In the event, the British Government put a motor car at the disposal of the official and had him accompanied to Islay. He wrote a report which, according to press leaks, recommended that infringement proceedings be started against the UK in the European Court. But as so often happens, the matter was settled before it reached the Court. The UK Government's Scottish Office persuaded the distillery to think again. The distillery decided that another source of peat was suitable. The digging of Duich Moss did not proceed.

So here we have a local dispute decided first one way and then another. Originally it was an aspect of home affairs, but by the time it was finished it had become something different. But it certainly was not foreign affairs in the traditional sense of the relations between two or more sovereign states, and one would have to look hard to find an international treaty governing relations between two nation states which allowed an official of one to travel to another so as to overturn a locally made decision.

Of course the outcome was not just of local interest since not only did it involve a site of importance for a significant proportion of the world population of a protected species but it also had implications for other Member States when dealing with the birds Directive. British home policy in this case can be criticised, not just for a failure to identify a site as required by the Directive, but also for leaving to a local authority a decision that is really beyond it. In this case EC intervention has shown up a national error of judgement. An alternative view is that the local authority should itself have recognised the significance of the nature conservation aspect and of the EC Directive without central government involvement.

This story illustrates how EC policy is becoming so intertwined with home policy that it is not always possible to understand home policy without taking account of EC policy. Although EC policy has the attributes of foreign policy during its formulation, it then becomes integrated with home policy in its implementation. This originality was recognised by the European Court as long ago as 1964 when it described the EC as establishing 'a new legal order of international law' (ECJ 1963).

One can say that for those subjects where there is a corpus of EC law, the EC is best understood as a federal system with more than one legislature, but where the higher level of government is not itself a nation state. This is a reversal of the situation with existing federations with which we are familiar – the USA and Germany for example – in which the higher level is a nation state and lower level bodies are not.

Origins of EC environmental policy

Although the Treaty took effect in 1958, it was not until 1972 that the Heads of State and Government decided that the EC should adopt an environmental policy and called upon the Commission to draft a programme of action. The Treaty, being a creation of the 1950s, did not refer to protection of the environment. It was developments in the Member States, and in other international fora, such as the 1972 UN Conference on the Human Environment at Stockholm, that created the pressure for the EC to move in this direction. It was entire coincidence that Britain, Ireland and Denmark joined the EC in the very year that it adopted the first action programme on the environment.

It was not until 1987 when the Single European Act amended the Treaty of Rome that environmental policy was explicitly provided for. Until then environmental legislation was adopted under a rather generous interpretation of the Treaty that attracted criticism despite some endorsement by the European Court (House of Lords 1978).

The first action programme on the environment covered the period from 1973 to 1976 and we are now two thirds of the way through the fourth action programme. These programmes have two main purposes. They suggest specific proposals for legislation that the Commission intends to put forward over the next few years, and they provide an occasion to discuss some broad ideas in environmental policy and to suggest new directions for the future.

The first programme was needed to chart a wholly new course and was a long and comprehensive document. It started with a general statement of the objectives and principles of EC environmental policy and then went on to spell out action that the Commission would propose.

The international dimension of environmental policy was recognised from the beginning and no less than four of the eleven principles relate to international relations. They can be paraphrased as follows:

- Activities carried out in one country should not cause deterioration of the environment in another.
- The effects of environmental policy in the Member States must take account of the interests of the developing countries.

- The EC and the Member States should act together in international organisations and in promoting international and worldwide environmental policy.
- In each category of pollution it is necessary to establish the level of action (local, regional, national, EC, international) best suited to the type of pollution and to the geographical zone to be protected.

The Single European Act gave legal force to several of the eleven principles and added the most important new principle that 'environmental protection requirements shall be a component of the Community's other policies'. Concerning cooperation with other countries on environmental matters, the Single European Act used the words quoted at the head of this chapter. Before discussing what this means in practice, we must briefly look at the EC's internal environmental policy.

The EC's internal environmental policy

The EC has now adopted over 280 items of environmental legislation. Many of these are of a narrow technical character with little policy content, but several are important by any standard and some, such as that dealing with acid rain, for a time entered the realm of 'high politics'. They have all been described elsewhere (Haigh 1990) and only a few are touched on here.

Given that EC environmental policy began without a clear legal base in a Treaty that was primarily concerned with the creation of a common market, it is no surprise to find that much environmental legislation is concerned with setting standards for products in trade, or with avoiding distortions to competition in industrial activities. If it could be shown that standards set in just one country were affecting the common market then there was a justification for EC involvement. The role of environmental policy then was to ensure that the common standard was a high standard. But even from the beginning the EC concerned itself with matters that hardly touched the common market such as standards for bathing waters or bird protection. Protection of birds is a subject that obviously cannot be handled at national level alone since birds fly across frontiers, but the justification for EC as opposed to national standards for bathing water is harder to find. The EC never confined itself to the two classic justifications for international environmental policy, first, that some issues (air pollution, sea pollution, transfrontier rivers) are not confined by national frontiers and may indeed be global in character and, second, that international trade is impeded by differing standards.

The adoption of any item of EC legislation is a process of accommodation between a number of countries. Sometimes, such as with standards for drinking water, each individual Member State will be concerned largely with the effects on its own internal procedures and with the cost, since what other Member States do will have little or no impact on the environment of others. With other Directives environmental effects between countries have been at the heart of the discussions. An obvious case is the so-called Seveso Directive named after the suburb of Milan where a major accident at a chemical plant spread dioxin across the countryside.

Under pressure from the European Parliament who suspected that the Swiss company had located its plant in Italy because of laxer standards, the Commission proposed a Directive. Provisions of the Directive that was finally adopted require the production of a safety report, of an 'on-site' emergency plan, an 'off-site' emergency plan, and that the local population be informed of the correct behaviour to adopt in the event of an accident.

The proposal as put forward by the Commission had no transfrontier provisions but the Benelux countries pressed for a requirement that a Member State give all appropriate information to other Member States that might be affected by an accident. Member States would also have had the right to consult on the necessary measures. This was resisted by the French Government. Although the proposed Directive specifically excluded nuclear power stations, France feared that it would set a precedent which might then be used by other Member States to comment on the siting of French nuclear power stations. This dispute held up adoption of the Directive by eighteen months and in the end a compromise was achieved under which information is to be made available between Member States only 'within the framework of their bilateral relations'. This means that the Commission does not have the same right to insist on implementation as it would have had otherwise, and means that other Member States have no rights at all.

Another important Directive, known as the 'sixth amendment', requires the manufacturer of any new chemical to supply a file with the results of tests for effects on man and the environment to an authority (see Chapter 7). The authority can call for extra information and if not satisfied can prevent marketing of the chemical. The file is also sent to the Commission which passes it to the authorities in all Member States. Any one of these authorities can ask for further information. Unless objections are raised within 45 days, the manufacturer has assured access to the whole EC market. The Directive simultaneously seeks to prevent environmental problems arising and serves the purpose of a common market in chemicals. Its successful operation depends very much on collaboration and mutual trust between the authorities in different countries.

The 'sixth amendment' was adopted in 1979 with support from the chemical industry in Europe, which is surprising given that the industry was not then known for welcoming environmental legislation. One explanation for this is that comparable Regulations were being developed in the USA by the Environmental Protection Agency under the Toxic Substances Control Act. The European chemical industry feared that impediments could be put in the way of exports to the USA, and believed that comparable rules in Europe could be used as pressure on the US authorities. They felt that the USA might be deterred by denial of access to the whole EC market – which is considerably larger than the US market – in a way in which they would not be deterred by lack of access to, say, the French or German markets alone. In the event, no trade war across the Atlantic has developed, and the hope remains that ultimately the procedures that enable a new chemical to be sold in the US market will be recognised in Europe and vice versa.

The best known item of EC pollution legislation dealing with a transboundary issue is the Directive that seeks to tackle the issue of acid rain. Known as the large combustion plant Directive, it limits emissions of sulphur dioxide and nitrogen oxides from such plants (see Chapter 4).

The issue of acid rain reached the political agenda in Sweden in the late 1960s, and in 1972 Sweden managed to make it an international issue at the UN conference on the environment held in Stockholm that year. Pressure from Sweden and the other Nordic countries eventually led to the Geneva Convention on long-range transboundary air pollution being adopted in 1979 under the auspices of the UN Economic Commission for Europe. This UN grouping includes not only countries from Western and Eastern Europe, but also the USA and Canada. Largely because of opposition from the USA, West Germany and the UK, the Convention did not include any firm targets for reduction in emissions.

The key date for acid rain policy in Europe is 1982, when the German Government changed its position from passive resistance to enthusiastic support for a policy of significant reduction in emissions. It then persuaded the Commission to propose a Directive similar to German legislation. After a long struggle, a Directive was agreed that set emission standards for new plant and required each Member State to reduce emissions from existing plant by certain percentages in three stages. The Commission had originally proposed that each Member State should reduce emissions by 60 per cent by the year 1995 compared to a 1980 baseline. The economic, geographical and fuel supply circumstances were so different that a uniform reduction proved unacceptable to several countries. As a result a compromise was eventually achieved with different countries having quite different reductions that nevertheless should result in a 58 per cent reduction in overall EC sulphur dioxide emissions by the year 2003 – a slippage of two percentage points and eight years from the Commission's original proposal. This differentiated reduction may yet prove a better model for the global warming issue across the world than the uniform reductions agreed for ozone-depleting substances.

The EC's external environmental policy – conventions

The development of the EC's external powers – that is to say its ability to deal with other countries much as does a nation state – has come about largely as a result of decisions of the European Court of Justice. The development of these powers has been analysed extensively but only rarely have their practical implications in the environmental field been discussed, very probably because they are well known only to those few officials involved. A rare glimpse of problems that have arisen is that given by Andre Nollkaemper (Nollkaemper 1987) drawing on interviews conducted during a course of probation in the Ministry of Foreign Affairs in the Netherlands. This account draws heavily on Nollkaemper's description. Needless to say, not all the problems that have arisen have been resolved, a major one being implicit in the quotation at the head of this paper.

Before the judgement of the European Court in the AETR case of 1971 (ECJ 1971) it was possible to argue that the Community was only competent to conclude international conventions when this was expressly provided for in the Treaty. Even this assumed that other countries were prepared to deal with the EC, and not all were. In the AETR case, and on a number of occasions since, the Court decided that competence for external affairs can also be implied by the Treaty as well as by the acts of the institutions performed under the Treaty. Of paramount importance is the link between internal and external powers. In the AETR case the Court held that whenever the EC has promulgated internal rules in a certain field, i.e. has taken measures binding on the Member States, then the powers to act externally in that field are created. Moreover, under certain conditions, these powers will be of an exclusive nature: as soon as the EC comes into possession of these powers, the Member States will have lost them. Needless to say, one of the most important – and complex – questions that arises concerns the conditions under which powers become exclusive.

A most important consequence is that the EC's external powers expand without the express approval of the Member States simply in the course of developing the EC's internal policies. An extra constraint has therefore been added to EC internal policy-making since the Member States should now always consider whether the adoption of some desirable item of EC legislation might not result in the undesirable (to them) loss of external competence.

Only rarely is this soul searching made public but this happened in Britain during the negotiations that led to Directive 80/51 on aircraft noise. In a debate in the House of Commons concern was expressed that adoption of the Directive might lead to an extension of Community competence into the field of aviation generally and not just to aircraft noise. Some felt that the Commission might want to represent the Member States at international meetings, such as those of the International Civil Aviation Organization (ICAO). However, any fears in the minds of the UK Government on that occasion seem to have been overcome since the Minister, Norman Tebbit, expressed himself satisfied that:

> ... the extension of the Community's authority which the Directive will produce is justified by the extra powers that it will bring to enable us to limit the noise of aircraft from other Member States
>
> *(House of Commons 1979)*

These words suggest a balancing of the risks and advantages. The reason for resisting EC competence in this area was given by the Minister. He explained that 'political issues were rarely raised in ICAO' and 'the power blocs are little in evidence'. The implication was that if the EC began to act as a bloc this would encourage other countries to do likewise. The UK Government must have felt confident that other Member States would think in the same way, and it is interesting to note that although three years later the Commission asked for observer status at ICAO this was refused by the Member States.

Another example is the refusal by the Council to adopt a Directive on the dumping of wastes at sea, which the Commission had put forward at least partly in order to be able to accede to international dumping conventions (the Oslo and London Conventions). Some Member States believe that accession by the EC will lead to a duplication of activities and that accession by the EC will add nothing to environmental protection.

Where the EC has exclusive competence for all the subject matter of a Convention, it is possible for the EC to become a party to it without the Member States also being parties. This has happened in some fields, such as trade in commodities, but it has not yet happened in the environmental field. Conventions to which some or all of the Member States have become parties as well as the EC are known as 'mixed agreements'. A recent example of the consequences of this arose with the Montreal Protocol to the Vienna Convention for the Protection of the Ozone Layer. At the meeting in London in June 1990 at which amendments to the Protocol were agreed, the EC had competence for negotiating the percentage reductions in the quantity of ozone-depleting substances that could be produced because of the existence of an EC Regulation covering this. However, the EC had no competence for the decision to establish a fund to assist third world countries to obtain the more expensive alternative substances and technologies, and on this point the Member States acted on their own.

Where, under a mixed agreement, the EC has competence the Commission will negotiate on behalf of the EC in accordance with a mandate given unanimously by the Council, i.e. by the Member States. Sometimes this is set in advance but sometimes instructions are given *sur place*. Where there is no unanimity among the Member States problems arise. In 1985 at a meeting in Buenos Aires of the parties of the Washington Convention on international trade in endangered species (CITES) unanimity could not be found so that all ten Member States abstained, leading to a substantial loss of influence.[6] In the same year at a meeting of the parties of the Bonn Convention on the conservation of migratory species of wild animals agreement could not always be reached on matters where the EC had competence and it was accepted that Member States could not act independently. However, where the EC did not have exclusive competence it was decided that Member States could act independently. This means that individual Member States can advance different arguments and may even vote against each other, though this of course results in a breach of the principle of EC solidarity.

Yet another example of the problems of mixed agreements arose with the Paris Convention for the prevention of marine pollution from land-based sources. In 1982 the EC Commission had taken the view that nothing prevented Member States adopting more stringent standards under the Paris Convention than comparable EC standards, but since 1984 this has been disputed. Disagreements arose relating to standards for mercury, cadmium and PCBs [polychlorinated biphenyls] and the EC Commission even succeeded in preventing the adoption of proposals for PCBs on the grounds of the EC's exclusive competence. Needless to say, these disputes between the EC Commission and the Member States as to their relative powers are particularly irritating to non-Member States and may well be incomprehensible to them. They can paralyse the work under the convention. In the case of this Convention (following the accession of Spain and Portugal to the EC) the only non-Member States are Sweden and Norway. Sweden followed this episode by tabling a list of questions relating to the powers of the EC.

The EC is now a party to a large number of conventions covering a wide range of environmental concerns. It is seeking to be a party to more. The broad interest of the Commission, with the support of the European Parliament, is to enhance the position of the EC in the world. The extension by the Court of the EC's external powers has allowed this, but without recognition of the EC by other countries as an appropriate party to conventions this would have counted for nothing. As recently as the early 1980s during the negotiations of the Vienna Convention on the ozone layer, the USA was querying the basis on which the EC could participate. Eventually, with the support of the Soviet Union, and despite resistance from the Commission, the USA insisted on a clause in the Convention relating to the conditions under which the EC could be a party. Before that, in the 1970s, EC participation in the Helsinki Convention on the Protection of the Marine Environment of the Baltic Sea Area was refused by the other parties led by the Soviet Union.

The position with respect to third countries now seems to be that the principle of participation by a 'regional economic integration organisation' is accepted but the conditions, including voting procedures in the subsequent implementation of the Convention, still have to be negotiated each time. The precedent set in the Barcelona Convention (see below) is that when the EC is given voting powers and votes in the place of the Member States it has as many votes as the number of participating Member States. However, in the Baltic Fisheries Convention, which is not a 'mixed agreement', the EC only has one vote, although two Member States have a coastline on the Baltic. The more the EC acquires exclusive competence and the more it appears to third countries as if it is acting as a nation state itself, the greater will be the pressure for it to be accorded only one vote.

The participation clause

The evolution of the 'participation clause' in various conventions has been described elsewhere (Kiss and Brusasco-Mackenzie 1989). The first environmental convention to which the EC became a party (in 1975) is the Paris Convention for the prevention of marine pollution from land-based sources. This took place apparently without difficulty perhaps because a majority of the parties were EC Member States. Next is the Barcelona Convention of 1976 for the protection of the Mediterranean Sea against pollution, where Member States were in a minority among the contracting parties. This Convention includes Article 24 which mentions the European Economic Community by name and provides for it to sign as well as any other 'similar regional economic grouping at least one member of which is a coastal State of the Mediterranean Sea area and which exercises competence in fields covered by this Convention'. This created the theoretical possibility that other groupings, e.g. Arab League, Organisation of African Unity, could sign if, and to the extent that, sovereignty over the subject matter of the Convention had been transferred to them.

The Geneva Convention of 1979 on long-range transboundary air pollution drafted under the auspices of the United Nations Economic Commission for Europe (UNECE) does not name the EC but foresaw participation by 'regional economic integration organisations' constituted by sovereign Member States of the UNECE having competence to negotiate, conclude and apply international agreements. This formulation was accepted by the countries of Eastern Europe who did not at that time recognise the EC.

The EC has now clearly established itself as an actor on the international stage in environmental as well as in other affairs. Although the Commission is anxious to expand this role, we have seen that there have been subjects such as aircraft noise and dumping at sea where the Member States have, for certain reasons, prevented the EC adopting the role that it has acquired for other subjects. Even where the EC has become a party to international conventions the problems of competence arise. The extent of EC competence depends on internal EC rules, and because these may not be coterminous with those in a convention – and the provisions of a subsequent protocol to the convention may not be knowable in advance – the Member States are likely to retain some competence. The division of competences can be very confusing for third countries and can make negotiations difficult or even in the extreme can block them.

This complicated process has given the EC greater power in the world, but that by itself cannot be a justification. EC involvement on the international stage must be justified in the end by the extra contribution that the EC can make to solutions to international problems. It must add something to what the Member States acting independently could themselves have done. Fortunately there is an example in the environmental field.

The ozone layer – the Montreal Protocol

The hypothesis that certain gases called chlorofluorocarbons (CFCs) would deplete the stratospheric ozone layer was first advanced by scientists in 1974. Any decrease in the ozone layer allows more ultraviolet radiation to reach the Earth's surface and so increases the risk of skin cancers. The response to this hypothesis differed in different countries in the 1970s. The USA banned the use of CFCs in aerosol cans for non-essential uses, but did not regulate other uses, e.g. as solvents, refrigerants and in blowing foam. Canada, Norway and Sweden followed the US example. The EC took a different course and in 1978 the Council adopted a Resolution calling for a limitation on CFC production. Then in 1980 it adopted a Decision that placed a production capacity limit on two types of CFC. While the US action resulted in a significant reduction in production, the EC action had little immediate effect as EC production capacity was higher than actual production.

In 1977 the United Nations Environment Programme (UNEP) began a review of scientific aspects and in 1981 initiated negotiations for a global convention. The Council of the EC authorised the Commission to participate on behalf of the EC in these negotiations and in 1985 many countries, among them several EC Member States, as well as the EC Commission, signed the Vienna Convention for the Protection of the Ozone Layer. It is what is sometimes called a framework convention since it covers such matters as cooperation on monitoring and research but does not itself place any obligation on the parties to take any specific measures to protect the ozone layer. These were to be laid down in separate protocols.

During the negotiations a dispute broke out between two groups of countries – the EC and what was called the Toronto Group (Canada, USA, Finland, Norway and Sweden). Each group proposed that the first protocol to cover CFCs should embody the policies already adopted in their own group of countries. The Toronto Group's proposal was for a worldwide extension of a ban on uses of CFCs as aerosol propellants but involved no limit on other uses of CFCs. The EC, not surprisingly in view of the approach it had already adopted, proposed a production capacity limit. The Toronto Group advanced their proposal on the grounds that it was the quickest way of obtaining an immediate reduction in CFC releases. The EC maintained that an aerosol ban did nothing to prevent releases from growing non-aerosol uses and that, since it is the total amount of CFCs released that affects the ozone layer, the only effective action was to limit total production. As a result of this dispute, no protocol was adopted in 1985 and negotiations did not start again until 1986.

Before the new negotiations started the United States government changed its position. It dropped its proposed aerosol ban and proposed instead a freeze on CFC production by all countries followed by a series of reductions leading to a production ban. Effectively the USA had conceded the merit of the EC production limit approach though reformulated and extended in a much more stringent form. Arguably the log jam was broken when, first, US environmental organisations, and then industry, abandoned the US Government's original negotiating position and embraced the EC approach. While the EC's 1980 Decision was originally largely symbolic it had defined an intellectually defensible approach which ultimately became incorporated into the Montreal Protocol.

Following the US proposal, the EC in March 1987 agreed negotiating guidelines for the Commission which included a freeze at 1986 levels on entry into force of the Protocol followed by a 20 per cent reduction four years later. This was not achieved without considerable initial resistance from some Member States, including the UK, under the influence of their industries. In subsequent negotiations the EC agreed to a further cut amounting to a 50 per cent reduction by the turn of the century. This was embodied in the Montreal Protocol in September 1987, which came into force on 1 January 1989. The EC and the Member States ratified it simultaneously.

No sooner was the Protocol agreed than a consensus developed that the recently discovered hole in the ozone layer was caused by CFCs and it became evident that the reductions in the Protocol were not enough. Fortunately the Protocol included a review mechanism, and in June 1990 an amendment required additional reductions. In December 1990 the EC went further than the Protocol requires by agreeing to phase out CFCs by 1997.

The US deserves the credit for creating the pressure in 1986 and 1987 for significant reductions in CFC production and US negotiators did not always conceal their irritation with the EC for what they saw as foot dragging and the complications that it introduced (Benedict 1989). It is therefore worth speculating on what might have occurred had the EC not been involved. Presumably a protocol along the lines of the Toronto Group's proposal would have been adopted in 1985 and several EC Member States would no doubt have become parties. This would have been a less satisfactory protocol, which would have needed complete revision after the ozone hole discovery, and several important countries might well have stayed outside at least initially. The lack of solidarity would have weakened the whole effort. In the event, the EC not only ensured that the Protocol had a better form but also delivered intact a bloc of twelve industrialised countries central to any successful global action since between them they produced more CFCs than the USA or Japan or the USSR. The result was an ideal situation whereby several countries contributed solutions to a global issue and learned from one another during the process.

Global warming

The Vienna Convention with its associated Montreal Protocol is the first convention to deal with the global atmosphere and is therefore inevitably seen as a precedent for a possible convention on global warming. The idea of a freeze followed by percentage cuts applied equally to all countries is firmly before us.

In preparation for the Second World Climate Conference held in November 1990, the EC Council – at a meeting exceptionally composed of both environment and energy Ministers – agreed that the EC collectively should stabilise carbon dioxide (CO_2) emissions by the year 2000 at 1990 levels. No legally binding instrument was adopted but this 'political' decision enabled the EC Commissioner – Carlo Ripa di Meana – to make a bid for leadership. The Ministerial Declaration made at the end of the Conference welcomed 'the decisions and commitments undertaken by the European Community with its Member States' as well as those of a number of other countries to stabilise their emissions of CO_2. The United States was not one of these so it cannot be a foregone conclusion that this commitment will necessarily form part of a convention if the world's largest emitter of CO_2 is to be a party to it.

The EC's political decision was made possible by some Member States – and most notably Germany – having already agreed to cut their emissions significantly. This, combined with estimates of what might happen in other countries that had set themselves no targets, and the United Kingdom that had adopted a target of stabilisation by 2005, made stabilisation by 2000 for the EC as a whole a realistic possibility in the view of the Council. To translate this hope into more of a reality, the Commission could now propose a Directive allocating different targets to different countries on the model of the sulphur dioxide reductions set out in the large combustion plant Directive. Without some greater definition it is not at all clear what the overall EC commitment to stabilise actually means in practice to those Member States who have not yet adopted national targets.

Should a Directive along these lines be agreed, it would give the EC the necessary competence to become a party to a convention that involved stabilisation by the year 2000, or any subsequent date. But it cannot be assumed that the Council will agree such a Directive. We have already seen how the Council has on occasion not provided the Commission with a mandate to negotiate on its behalf on some subjects, and some Member States might not want to lose the power to negotiate on their own behalf on a subject that has such profound implications for national policies as does energy consumption. In this respect the Montreal Protocol does not provide a perfect analogy. CFCs are a traded product and for that reason the EC is almost bound to be involved in its regulation. The Member States by agreeing to the Montreal Protocol transferred competence to the EC for all further control over the quantity of CFC production, but since CFC production will now cease in a few years the loss of competence will soon be of theoretical interest only. CO_2 by contrast is not a manufactured and traded product but is the by-product of innumerable activities which will have to be controlled or influenced if agreed targets are to be achieved. Whole areas of national life including the pattern of industrial development, electricity generation, not to mention personal mobility will be affected forever by any target. Whereas individual countries may be willing to agree an initial target, they may well be apprehensive about the loss of influence over selecting subsequent and more stringent targets. One possibility which would enable the EC to be a party to a convention is for the EC to agree a Directive that sets an overall EC target of stabilisation by the year 2000 but instead of allocating individual differentiated targets for each Member State, requires each of them to produce an overall strategy for controlling CO_2 emissions involving detailed plans for all of the major sources of CO_2 and the practical steps for achieving these plans. The overall EC target would provide the pressure on the Member States to make their own contribution. If the aggregate of the national strategies is such that stabilisation by the EC would not be achieved, then pressure can be brought on those Member States who are not contributing enough.

The lesson of the Montreal Protocol can be applied here in reverse. Once the EC has adopted a policy as a result of a process of negotiation between twelve countries it becomes difficult to shift. This happened with CFCs and fortunately the policy the EC adopted in 1980 provided the model for aspects of the Protocol. But if it had been an inadequate policy it would not necessarily have been taken up in the Protocol. Similarly, if the EC adopts a detailed carbon dioxide policy on its own, which is not adapted to other countries, it may make it difficult for the EC and its Member States to play a constructive role in the negotiations on a global warming convention. Flexibility will therefore be needed.

The EC's competence in international environmental policy is suddenly being put to the test. Collectively the EC has already played a role and is exerting pressure, on the USA in particular, in a way that individually the Member States would not have been able to do. But it needs practical action on the part of the individual Member States if a reality is to be made of any target, be it stabilisation or reduction. Practical action could require flexibility that can be lost when power is lost. One can anticipate that the allocation of responsibility between the EC and its Member States as negotiations proceed over the years will require a consensus that may not be achieved without some struggle. The stakes are high and are worth the struggle. Other countries may sometimes have to show patience as they watch.

Published November 1990

Developments since 1990

The EU's ratification of the 1992 UN Framework Convention on Climate Change (FCCC) proved rather easier than the paper above suggested it might be. When signing the FCCC, the EU and its Member States made a formal declaration 'that the inclusion of the European Community as well as its Member States in the lists in the Annexes in the Convention is without prejudice to the division of competence and responsibilities between the Community and its Member States', thus making clear to other parties that the issue of their relative roles was not straightforward. The EU and its Member States ratified the FCCC in December 1993 after having adopted the 'monitoring mechanism' Decision 93/389 that enabled the EU and its Member States to fulfil their obligations collectively. The 'monitoring mechanism' Decision, as its name implies, took the softer of the two possibilities suggested above. It did not allocate different targets to different countries as did the 'large combustion plant' Directive (Chapter 4) but instead required Member States to devise their own national CO_2-reduction programmes and to submit these to the Commission. The Commission then evaluated these to see if the EU was on the path to meeting its commitment to stabilise CO_2 emissions at 1990 levels by 2000. If it looked as if collective stabilisation would not be achieved then the Member States and the Commission would have to decide among themselves what more would have to be done. This can be called a bottom-up, rather than a top-down approach. The story is taken forward in Chapter 9.

The EU has become a party to a number of environmental conventions since the paper above was written (Delreux 2011, 2013). While some of the uncertainties and ambiguities have been resolved, others remain. The questioning by other countries as to whether the EU should be a party to international conventions no longer happens, and the question of the number of votes allocated to the EU when it negotiates in place of the Member States no longer seems to be an issue.[7]

On the other hand, the underlying tensions between the Member States and the EU remain. The 2007 Lisbon Treaty prescribed certain rules about the relative roles of the Commission and Council when negotiating, signing and ratifying conventions, and although these have clarified some matters they cannot resolve all the tensions that are inherent in the very attempt to share sovereignty.

Notes

1 International 'treaties', 'conventions' and 'agreements' can all be equally binding. The name 'treaty' is usually reserved for subjects of the greatest importance. 'Agreement' is usually used for technical matters such as regulating trade in commodities. Most environmental matters have been the subject of 'conventions'.
2 The Council of Europe was created in 1949, a decade before the EU. Its hallmark has been the fostering of parliamentary democracy and human rights. It expelled Greece from membership during the 'Regime of the Colonels' in the 1960/70s. The European Court of Human Rights (ECHR) is one of its institutions. The ECHR is sometimes confused with the European Court of Justice (ECJ) which is an EU institution.
3 This is much less true today. The increased powers of the Parliament mean the process is more open, and some parts of Council meetings are now open.
4 The European Court of Justice has developed the doctrine of 'direct effect' whereby national courts can apply a Directive, in the absence of national implementing legislation, if its requirements are sufficiently clear.
5 There has been apparently no subsequent instance of a Commission official making a similar visit to a Member State to inspect a site and gather information at first hand although there is no rule to prevent it. Given the new emphasis on 'subsidiarity' (see Chapter 12), such a visit would perhaps now be regarded as too intrusive. The argument presented in this chapter is not affected by whether or not a site visit was made.
6 The Commission wanted abstention by the Member States in the absence of a common EC position being agreed. This was because, in its view, the issues under consideration (bullfrogs and hooded seals) were already within exclusive EC competence since CITES was a common commercial policy matter. Hence the Member States were not entitled to act unilaterally in an external context with respect to them. The Member States, or at least six of them, took a different view, arguing that, as the species were not already within the CITES Appendices or the EC CITES Regulation Annexes, they were not within EC competence and hence Member States were entitled, in the absence of an EC common position being established in accordance with Article 5 of the Treaty, to take whatever action they individually deemed best.
7 In a personal communication Dr Tom Delreux informed me that he had interviewed over 100 officials from the Commission and Member States who had conducted international environmental negotiations on behalf of the EU and none of them had mentioned issues with votes.

References

Benedict, R (1989) US environmental policy – relevance to Europe, *International Environmental Affairs*, 1(2), pp 91–102.

Delreux, T (2011) *The EU as International Environmental Negotiator*, Farnham: Ashgate Publishing.

Delreux, T (2013) The EU as an actor in global environmental politics. In Jordan, A and Adelle, C, eds, *Environmental Policy in the EU*, 3rd edn, London: Routledge.

ECJ (1963) Case 26/2 van Gend en Loos v. Netherlands Fiscal Administration

ECJ (1971) Case 22/70 Commission v. Council (AETR)

Haigh, N (1990) *EEC Environmental Policy and Britain*, 2nd edn, London: Longman.

Haigh, N (1992) The European Community and international environmental policy. In Hurrell, A and Kingsbury, B, eds, *The International Politics of the Environment*, Oxford: Clarendon Press.

Hurrell, A and Kingsbury, B, eds, (1992) *The International Politics of the Environment*, Oxford: Clarendon Press.

House of Commons (1979) *Official Report*, 19 June 1979, Cols 1251–1283.

House of Lords (1978) *Select Committee on the European Communities, Approximation of Laws under Article 100 of the EEC Treaty, 22nd Report, Session 1977–1978*, London: HMSO.

Kiss, A. and Brusasco-Mackenzie, M (1989) Les relations extérieurs de la CEE en matière du protection de l'environnement, *Annuaire Français de Droit International*, 35, pp 702–710.

Nollkaemper, A (1987) The European Community and international environmental cooperation: legal aspects of external community powers. In *Legal Issues of European Integration: Law Review of the Europa Institut, University of Amsterdam*, The Hague: Kluwer, pp 55–91.

3

SUSTAINABLE DEVELOPMENT IN THE EU TREATIES

To have a secure place on the EU stage, let alone to be at its centre, it was necessary for environmental policy to be embedded in the EU Treaties. That there had been no mention of the environment in the original treaty of Rome was a concern of both the European Environmental Bureau (EEB) and the Institute for European Environmental Policy (IEEP) when they were formed in 1974 and 1976 respectively. Both put forward ideas about treaty amendment and continued to follow the steps by which environmental ideas were introduced. The occasion to tell the story of how this happened presented itself when IEEP was part of an academic consortium that studied how several EU Member States were coming to grips with Agenda 21 – the name given to the framework for implementing sustainable development agreed at the UN Conference on Environment and Development (UNCED) held in Rio de Janeiro in June 1992. This resulted in a book called *The Transition to Sustainability* (O'Riordan and Voisey 1998) which included my chapter reprinted below (Haigh 1998).

The Brundtland Report (Brundtland Commission 1987) which defined and gave currency to the concept of 'sustainable development' was written in English, and in many languages it took time before there was generally accepted equivalent wording. During the study we found inconsistencies in how that phrase had been used in different language versions of the EU Treaties. Table 3.1, constructed by the participants in the consortium, shows four different German words used where 'sustainable' was used in English. We accordingly arranged for the table to be published in both English and German during the intergovernmental conference that drafted the Treaty of Amsterdam to influence the future choice of words. The Amsterdam Treaty signed in 1997 made 'sustainable development of economic activities' an objective of the EU. The Lisbon Treaty signed in 2007 now says the EU 'shall work for the sustainable development of Europe' and 'shall contribute to the sustainable development of the Earth'.

TABLE 3.1 The language of sustainable development in the Treaties and agreements of the European Community

	Rome Treaty Article 2	*Rome Treaty Article 130u (Objective of Development Cooperation)*	*Maastricht Treaty Article B*	*Agreement on the European Economic Area Preamble (1993)*
English	'sustainable and non-inflationary growth respecting the environment'	'the sustainable economic and social development of the developing countries'	'economic and social progress which is balanced and sustainable'	'the principle of sustainable development'
French	'une croissance durable et non inflationniste respectant l'environnement'	'le développement éconornique et social durable des pays en développement'	'un progres économique et social équilibre et durable'	'principe du développement durable'
German	'beständiges, nicht- inflationäres und umwelt- verträgliches Wachstum'	'nachhaltige, wirtschaftliche und soziale Entwicklung der Entwicklungsländer'	'ausgewogenen und dauerhaften wirtschaftlichen und sozialen Fortschritt'	'der Grundsatz der umweltverträglichen Entwicklung'
Dutch	'een duurzame en niet-inflatoire groei met inachtneming van het milieu'	'de duurzame economische en sociale ontwikkeling van de ontwikkelingslanden'	'een evenwichtige en duurzame economische en sociale vooruitgang'	'het beginsel van duurzame ontwikkeling'
Italian	'una crescita sostenibile non inflazionistica e che rispetti l'ambiente'	'lo sviluppo economico sociale sostenibile dei paesi in via di sviluppo'	'un progresso economico e sociale equilibrato e sostenibile'	'principio che lo sviluppo dev'essere sostenibile'
Portuguese	'urn crescimento sustentavel e não inflacionista que respeite o ambiente'	'o desenvolvimento economico e social sustentavel dos paises em vias de desenvolvimento'	'um progresso economico e social equilibrado e sustentavel'	'principio de urn desenvolvimento sustentavel'
Danish	'en baeredygtig og ikke-inflationaer vaekst, som respekterer miljoet'	'en baeredygtig efkonomisk og social udvikling i udviklingslandene og swrlig i de mest ugunstig stillede blandt disse'	'at fremme afbalancerede og vafge okonomiske og sociale fremskridt'	'princippe om baeredygtig udvikling'
Spanish	'un crecimiento sostenible y no inflacionista que respete el medio ambiente'	'el desarrollo económico y social duradero de los paises en desarrollo y particularmente, de los mas desfavorecidos'	'un progreso económico y social equilibrado y sostenible'	'del principio del desarrollo sostenible'

	Rome Treaty Article 2	*Rome Treaty Article 130u (Objective of Development Cooperation)*	*Maastricht Treaty Article B*	*Agreement on the European Economic Area Preamble (1993)*
Greek*	'synexi ke isorropi epektasi tis economies'	'ti statheri ke diarki economiki ke kinoniki anaptyksi ton anaptysomenon horon'	'isorropi ke statheri economiki ke kinoniki proodo'	'tin arxi tis viosimis anaptyksis'
Swedish	'hallbar och ickeinflatorisk tillvaxt som tar hansyn till miljon'	'varaktig ekonomisk och social utveckling i utvecklingslanderna'	'framja valavvagda och varaktiga ekonomiska och sociala framsteg'	'principen om en varaktig utveckling'
Finnish	'ymparistoa arvossa pitavaa kestavaa kasvua, joka ei edista rahan arvon alenemista'	'kehitysmaiden kestavad taloudellista ja sosiaalista kehirysta'	'edistaa tasapainoista ja kestavaa taloudellista ja sosiaalista edistysta luomalla alueen'	'erityisasti noudattaen kastavan kehityksen seka ennalta varautuvien'
Irish	'les fas inbhuanaithe neamhbhoilscitheach a urramaionn an comhshaol'	'le forbairt inbhuanaithe eacnamaich agus soisialta na dtfortha I mbeal forbatha'	dul chun chin eacnamaioch agus soisialta ata cothromuil inbhuanaithe a chur ar aghaidh'	'ar bhonn phrionsabal na forbartha inbhuanaithe'
Non-Member State (applied for membership – but membership rejected in a referendum)				
Norwegian	'en baerekraftig ikke-inflasjonsdrivende vekst som tar hensyn til miljoet'	'en baerekraftig okonomisk og sosial utvikling i utviklingslandene'	'A fremme likevektig og \Twig okonomisk og social framgang'	'prinsippet om en baerekraftig utvikling'

* The text provided in the table is a transliteration of the Greek script.

Introducing the concept of sustainable development into the Treaties of the European Union – book chapter published in 1998

The Treaty of Rome that established the European Economic Community in 1958 made no mention of the environment nor did it suggest that there were any limits to the 'continuous and balanced expansion' that was to result from establishing a common market and approximating national economic policies. This was to be the task of the Community as set out in Article 2 of the Treaty.

While the Treaty was truly original in creating an entirely new kind of 'Community' in which sovereign states ceded legislative powers in certain fields in order to achieve objectives that were beyond the reach of them individually, it remained a creature of its time in its call for 'continuous expansion' without any recognition that in a finite world environmental considerations must impose some limits. Not till some 30 years later was the Treaty amended to provide a legal base for environmental policy. Even then, the amendment made by the Single European Act in 1987 left untouched the language of the 1950s in which the task of the Community was defined. By coincidence the amendment to the Treaty was made in the same year that the Brundtland Commission gave currency to the concept of sustainable development, which recognised that the needs of future generations must not be compromised by the type of development pursued today. The Brundtland concept was quickly endorsed by many governments. When, therefore, some three years later, an intergovernmental conference began negotiations on what eventually became the Maastricht Treaty, the conference was receptive to the idea of amending the call for continuous growth in Article 2. Unfortunately, for a reason that we explain below, the wording adopted in the amended Article 2 was not ideal, so that the intergovernmental conference that began reconsidering the Treaty in 1996 decided to look at the subject again. The resulting draft Treaty of Amsterdam – which had yet to be ratified at the time of writing – finally introduced the word 'sustainable development' but alongside 'sustainable growth'.

This chapter traces the introduction of the concept of sustainability in the European Community and its Treaties, a process that can conveniently be divided into the following four periods:

- 1958–1972 – the dark ages
- 1972–1987 – environmental policy is established
- 1987–1991 – after the Single European Act
- 1991–1997 – from Maastricht to Amsterdam.

The dark ages (1958–1972)

During the first 15 years of the Community's existence there was nothing that could be called an environmental policy. The task of the Community, set out in Article 2, was as follows:

> The Community shall have as its task, by establishing a common market and progressively approximating the economic policies of Member States, to promote throughout the Community a harmonious development of economic activities, a continuous and balanced expansion, an increase in stability, an accelerated raising of the standard of living and closer relations between the States belonging to it.

While a few items of legislation which we would now call environmental were adopted during those 15 years they were always secondary to some other purpose. In 1967 a Directive was adopted on classifying dangerous substances, and their appropriate packaging and labelling, but the driving force was the desire not to allow the creation of barriers to trade by differing national rules set to protect workers. The same applies to the first Directives adopted in 1970 setting standards for emissions and noise from vehicles. While individual Member States may have seen the need for some environmental protection measures, the role of the Community was then only to prevent these undermining the 'common market'.

One subject which the Community did recognise from the beginning as requiring environmental standards was ionising radiation. The Euratom Treaty, signed in Rome in 1957 at the same time as the better known Treaty of Rome, was intended:

> ... to contribute to the raising of the standard of living in the Member States ... by creating the conditions necessary for the speedy establishment and growth of nuclear industries
>
> *(Article 1)*

The next Article then went on to say that in order to do this the Community should 'establish uniform safety standards to protect the health of workers and of the general public and ensure that they are applied'. The first Euratom standards relating to ionising radiation were accordingly laid down as early as 1959 but here again they were ancillary to the main purpose of advancing the nuclear industry, then optimistically seen, in the aftermath of Hiroshima, as turning swords into ploughshares.

Towards the end of this period (1958–1972) the Community was placed in a dilemma by the worldwide movement of thought that began to recognise that environmental protection was not just a minor or local matter but that the very future of human life on the planet was in issue. It was coming to be seen not just as an adjunct to other policies but as a subject that deserved attention in its own right and one requiring other policies to be reshaped. Several Member States at that time established Ministries for the environment (e.g. both France and Britain did so in 1970) and the subject was firmly placed on the agenda by the great UN Conference on the Human Environment held in Stockholm in 1972. The Community had to decide either that it should not involve itself in this new found subject, since it was not provided for in the Treaties, or that it would have to respond in some way.

Environment policy is established (1972–1987)

The Community had originally been founded with six Member States (France, Germany, Italy, Belgium, the Netherlands and Luxembourg). The first enlargement took place in 1973 with the accession of the UK, Ireland and Denmark (Norway had applied but rejected membership in a referendum). During the discussions leading to enlargement the Community replied to the criticism that it was too concentrated on a common market – that it was a 'businessman's club' – by deciding to expand its activities in three fields: consumer protection, regional policy and environment policy. This was described at the time as giving to the Community a 'more human face'.

At a Summit Meeting in Paris in 1972 the heads of State and Government, including those from the applicant countries, declared that the Community would embark on an environment policy in these words:

> … economic expansion is not an end in itself: its first aim should be to enable disparities in living conditions to be reduced … It should result in an improvement in the quality of life as well as in standards of living. As befits the genius of Europe, particular attention will be given to intangible values and to protecting the environment so that progress may really be put at the service of mankind.

This Declaration neatly sidestepped the issue of whether it was necessary to amend Article 2 before the Community could embark on this new policy. The Declaration effectively accepted that the existing Treaty was flexible enough to allow for environmental policy without having to be modified. The Declaration can, according to taste, be regarded as either a denial of the impossibility of continuous expansion in a finite world or a redefinition of the quality of that expansion. There was an implication that if 'progress' in the past had been pursued for its own sake and not for the service of mankind, this could now be changed.

The Declaration called on the Commission to draft an 'Action Programme on the Environment' and the first was published in 1973. When approving the Action Programme the Council also indulged in sleight of hand. They declared that the task of the Community set out in Article 2 'cannot now be imagined in the absence of an effective campaign to combat pollution and nuisances or of an improvement in the quality of life and the protection of the environment'. The use of the word 'now' is revealing: although the Treaty had not changed society's views apparently had. The Action Programme resulted in an explosion in the number of items of EC environmental legislation over the next 15 years, and while many were rather narrow and technical some are of major importance by any standard and have come to influence all the Member States.

From time to time during this period attempts were made to suppress environmental policy as being a distraction from the Community's main purpose. The hostility that environmental policy encountered may be difficult to understand today as was resistance at that time to Treaty amendment. But environmental policy survived and managed gradually to consolidate itself. Nevertheless doubts continued to be raised about its legality (House of Lords 1978). All items were then based on either Article 100 or Article 235. Article 100 provides for an approximation of laws among Member States so that trade is not distorted, and Article 235 empowers the Community to take measures to deal with any unforeseen circumstances which would otherwise impede the objectives of the Community – but as we have noted environmental protection is not one of these mentioned in the Treaty. Various pronouncements of the European Court of Justice gave some reassurance, such as the statement in 1983 that 'protection of the environment is one of the essential objectives of the Community which may justify certain limitations on the principle of the free movement of goods' – despite the lack of any authority in the Treaty for such an assertion (ECJ 1983). Whatever the reassurances, pressure for a clear legal base for environmental legislation continued to come from lawyers, of course; from policy institutes and parliaments (von Moltke 1977; House of Lords 1980); and also from the German Länder who were concerned at the uncertainty over the loss of their powers whenever Community legislation was negotiated by their Federal Government.

Simultaneously pressure for amendment of Article 2 to redirect the priorities of the Community came from environmental pressure groups. For example, the European Environmental Bureau (EEB) in its 'Manifesto' for the first direct elections to the European Parliament in 1979 had advanced the concept of sustainability and called for Europe to invent 'a new industrialism, a mature pattern of growth, that will enable its own great population to live more fully, yet press more lightly on the planet' (EEB 1979). The Manifesto ended with a call for amendment of Article 2.

It is probable that the first calls to remove 'continuous expansion' from Article 2 were made in Britain, possibly because accession to the Community had led many to read the Treaty closely. The Civic Trust in 1974 published an article (Civic Trust 1974) called 'Growth limits and the Treaty of Rome' proposing amendment of Article 2, and the Conservation Society proposed a form of words for a new Article 2 in 1975 (quoted below). These bodies were both members of the EEB which at its first meeting in December 1974 decided that its primary aim was 'to work for a sustainable life style in the European Community'. This was then written into its constitution providing an early example of the use of the Brundtland concept in a developed world context. Previously the phrase 'sustainable development' had only been applied to developing countries.

Three key ideas about inserting the environment into the Treaty had developed by the end of the 1970s: the need for a clear legal base for environmental legislation; the need to replace 'continuous expansion' in Article 2 with wording that showed respect for environmental quality and the needs of future generations (sustainability); and an obligation that all Community policies should take account of the environment (integration). These three ideas were rehearsed in 1979 in evidence before a House of Lords Committee (House of Lords 1980), and had already in 1977 been elaborated in an article by Konrad von Moltke, founding Director of the IEEP (von Moltke 1977). Indeed, the competence of the Community in the field of the environment was one of the first themes that IEEP tackled when it was formed in 1976. Several relevant publications are listed in von

Moltke's 1977 article, including a paper by Edgar Faure, President of IEEP and a former French Prime Minister.

Calls for a restructuring of the Community in different ways were also coming from other quarters. In 1984 the European Parliament adopted its draft Act of European Union (Commission of the European Communities 1984) inspired by the Spinelli report which foresaw the Community developing along the lines of a federal state with the Commission being elevated to the status of parliamentary government, and the Council and Parliament being co-equal partners in a bi-cameral legislature. Encouraged by proposals from environmental bodies, the Parliament's 'Draft Act' included a Title on environmental policy.

Meanwhile, certain governments had been pressing for faster progress in one of the Community's original tasks, a process that came to be known as 'completion of the internal market'. This resulted in the Commission publishing a White Paper on the subject in early 1985. In the ensuing discussion it became evident that in order to take several difficult decisions necessary to complete the internal market the power of individual Member States to veto decisions would have to be reduced. Amendment of the Treaty was therefore necessary and this provided the opportunity for the Parliament's 'Draft Act' to be taken forward at an intergovernmental conference. This eventually produced the Treaty known as the 'Single European Act' that *inter alia* amended the Treaty of Rome. The agitation of environmental bodies and others had been transmitted via the European Parliament and resulted in a new Treaty Title on the environment which we now discuss.

After the Single European Act (1987–1991)

The Single European Act marked the coming of age of Community environmental policy by more than just confirming its legitimacy. The stated objectives of environmental policy were broad and allowed for subjects that could well have been excluded even under the elastic interpretation of the Treaty that had previously prevailed. One example is the Directive on freedom of access to information on the environment.

The new Title furthermore set out certain principles of Community environmental policy which, although already in the action programmes, gave them greater authority. It then added the important new principle that 'environmental protection requirements shall be a component of the Community's other policies'. This requirement for integration is arguably the most important feature of the new Title since for the first time it gave authority to those concerned with environmental matters to question those developing the Community's other policies such as agriculture and transport. No longer was Community environmental policy just a matter of legislation covering pollution of water and air and nature protection which might well cause pain in the Member States but did not much touch other Directorates-General of the Commission: now it had the capacity to influence the policies of other Directorates-General. Increasingly it became a matter of dispute between them and so raised environmental protection to a higher political level. The new integration requirement is essential in shifting the Community in the direction of sustainable development.

The Single European Act made another important innovation relating to environment policy. It introduced 'qualified majority voting' for the adoption of legislation relating to the internal market together with the associated 'co-operation procedure' with the European Parliament. Although most environmental legislation continued to require unanimity in the Council, items relating to traded products, for example, could now be agreed by qualified majority voting. The best known example is the emission standard for cars that effectively made catalytic convertors compulsory (see Chapter 10). Several Member States were opposed to the standard which would never have been agreed by unanimity. Despite these significant changes, which embodied two of the three ideas developed in the 1970s, Article 2 was left intact with its original 1950s wording so that pressure for amendment to that Article continued.

One of the difficulties was that during the late 1970s and early 1980s no form of words with some chance of winning broad political support had been put forward as an alternative to 'continuous expansion'. The European Environmental Bureau suggested none in its Manifesto for the 1979 elections that called for amendment of Article 2 even if one of its members, the Conservation Society, had proposed the following words for Article 2:

> the highest quality of living conditions for all the people within the Community that are consonant with the paramount need, having regard to the interests of succeeding generations, to conserve the natural resources and the environment ...[1]

This formulation which approximates quite closely to the Brundtland definition of sustainable development (quoted below) had the disadvantage that it was rather long winded.

By coincidence it was in the same year that the Single European Act came into force (1987), that the Brundtland Commission published the report giving currency to the phrase 'sustainable development' and providing the well-known definition: 'development that meets the needs of the present without compromising the ability of future generations to meet their own needs'. Rather surprisingly the Brundtland report was quickly endorsed by a number of governments so that when in 1990 another intergovernmental conference began to negotiate what was to become the Treaty of Maastricht, the conference was receptive to the idea of replacing 'continuous expansion' with 'sustainable development'. Unfortunately there were two parallel intergovernmental conferences, one on monetary union and one on political union. While the conference on political union apparently found no difficulty with 'sustainable development', the conference on monetary union wanted wording in Article 2 calling for 'sustained non-inflationary growth'. The draughtsman who had to reconcile the work of the two conferences produced the following English language formulation that appears in the ratified version of the Maastricht Treaty. While the intention to introduce the Brundtland concept certainly underlies the wording, it is arguable that the combination with another, quite different, intention has resulted in wording that is either meaningless or at best ambiguous:

> The Community shall have as its task, by establishing a common market and an economic and monetary union and by implementing the common policies or activities referred to in Articles 3 and 3a, to promote throughout the Community a harmonious and balanced development of economic activities, sustainable and non-inflationary growth respecting the environment, a high degree of convergence of economic performance, a high level of employment and of social protection, the raising of the standard of living and quality of life, and economic and social cohesion and solidarity among Member States.

The Maastricht Treaty also made three other changes that are significant to environmental policy and sustainability (Wilkinson 1992). It stated that Community policy should be based on the precautionary principle (see Chapter 13); it strengthened the integration requirement; and it made qualified majority voting the standard procedure for environmental legislation subject to only a few exceptions.

By the time the Commission proposed the fifth Action Programme on the Environment in 1992 it felt able to call it 'Towards Sustainability'. To a large extent it was concerned to ensure that environmental considerations, including those of the long term, were integrated into a number of other Community policies.

From Maastricht to Amsterdam (1991–1997)

The intergovernmental conferences that drafted the Maastricht Treaty recognised that some matters had been left unresolved and called for a new intergovernmental conference (IGC) to meet in 1996 to deal with unfinished business. Since one of the tasks of the new IGC was to fit the Community for the challenges of the twenty-first century it is no surprise to find that environmental bodies again argued that the wording of Article 2 should be improved (IEEP 1995; EEB 1995).

The task of finding an appropriate form of words and having it agreed by 15 countries would be difficult enough if the Community conducted its business in only one official language, but there are 12 official languages. In not all of these is there a widely accepted equivalent of the English words 'sustainable development'. This is shown in Table 3.1 that sets out the use of the word 'sustainable' in all official languages of the Community in the following three separate Treaties:

- the Treaty establishing the European Community (the Treaty of Rome) (as amended by the Treaty of Maastricht but before the Amsterdam amendments);
- the Treaty on European Union (the Treaty of Maastricht); and
- the Agreement on the European Economic Area which came into force in 1994 and governs relations between the European Community and certain European countries that are not part of the European Community.

In some languages the same word is used in all four columns of the table, while in others different words are used. An extreme case is provided by the German texts in which a different word is used for the English 'sustainable' in all four columns. In particular in the important Article 2 the word *beständig* is used. This can be translated as continual, continuous, permanent, lasting and stable. The word has not elsewhere been used as a translation of 'sustainable' in the sense defined by the Brundtland Commission, and as used in Article 2 the context suggests the traditional economic sense of continuous growth. Thus in German the Maastricht version of Article 2 is little different in this respect from the 1950s original.

Table 3.1 was published during the IGC to stimulate debate in different countries about the need for appropriate language (Haigh 1996; Haigh and Kraemer 1996).

The ideas put forward in early 1995 by environmental bodies for improving the Treaty included clarifying the wording on sustainable development and revising appropriate articles on specific policies such as transport and agriculture to require environmental considerations to be taken fully into account. These ideas caught the attention of Ministers, or reinforced ideas that were already there, because one year later, just before the beginning of the intergovernmental conference (IGC), a number of Member States announced that they were seeking environmental changes. The European Commission and the European Parliament said the same. The countries concerned were mostly the smaller ones: Sweden, Denmark, Finland, Austria, the Netherlands, Belgium and Luxembourg. Some said they wanted the concept of sustainable development stated as an objective of the Treaty and also wished to see a reinforcement of the requirement to integrate the environment into other policies. Sweden in particular stated that its government 'intends to propose that an environmental goal for the Common Agricultural Policy is introduced into Article 39 of the Treaty of Rome'.

When in March 1996 the European Council agreed the agenda for the IGC the environment was one of the items listed, and the Italian Presidency's conclusions expressed it in these terms:

> A healthy and sustainable environment is also of great concern to our citizens. Ensuring a better environment is a fundamental challenge for the Union. The IGC will have to consider how to make environmental protection more effective and coherent at the level of the Union, with a view to a sustainable development.

Nine months later in December 1996 the Irish Presidency issued a general outline for a draft version of the Treaties based on the work already undertaken by the IGC. This amended Article 2 to introduce 'sustainable development' and a new Article 3d (now 3c) which strengthened the integration requirement 'with a view to promoting sustainable development'. The Irish text of Article 2 was agreed virtually unchanged at Amsterdam in June 1997 so that the third idea developed in the 1970s had finally entered the Treaty – subject to successful ratification.

The new Article 2 reads:

> The Community shall have as its task, by establishing a common market and an economic and monetary union and by implementing the common policies or activities referred to on Articles 3 and 3a, to promote throughout the

> Community a harmonious, balanced and sustainable development of economic activities, sustainable and non-inflationary growth, a high degree of convergence of economic performance, a high level of employment and of social protection, a high level of protection and improvement of the quality of the environment, and the raising of the standard of living and quality of life, and economic and social cohesion and solidarity among Member States.

The new Article 2, unlike the Maastricht version, separates 'balanced and sustainable development of economic activities' from 'sustainable and non-inflationary growth' and also adds 'a high level of protection and improvement of the quality of the environment'. It is not immediately clear that 'sustainable development' is to take precedence over 'sustainable growth', nor what is meant by 'sustainable growth'. However, if the Brundtland definition of 'sustainable development' is followed then 'sustainable growth' cannot mean ever increasing consumption of limited resources since it would not then meet 'the needs of the present without compromising the ability of future generations to meet their own needs'. It would have been clearer if the needs of future generations had been introduced explicitly. Nor is 'sustainable development' linked to environmental protection though the link is made by the new Article 3c which reads:

> Environmental protection requirements must be integrated into the definition and implementation of Community policies and activities referred to in Article 3, in particular with a view to promoting sustainable development.

This brief new Article 3c does four other things. It makes clear that 'integration' is intended to result in sustainable development. It applies to all policies and activities in Article 3, which is more precise than the existing wording which just says 'other Community policies'. 'Sustainable development', having been connected to environmental protection, is made to run through all policies. It strengthens the integration requirement politically, and probably legally, by lifting it out of Article 130r and placing it more prominently. The result is that were it ever to be tested, the Court of Justice would be likely to say that the changed position was intended to have an effect, and would accord the environment greater importance. It is not impossible to imagine a Member State that is dissatisfied with an item of legislation for being insufficiently environmental challenging it for not fulfilling the requirements of Article 2 or Article 3c.

The Treaty of Amsterdam did not amend Articles dealing with specific subjects such as agriculture and transport, but if and when they come to be amended at another IGC they too can be brought up to date by introducing appropriate environmental objectives.

Has the struggle over many years to introduce appropriate environmental language been worth the effort? There are those who argue that changing words is just playing with words and that action is what matters. They overlook the fact that the wrong words make the right action more difficult and the right words make it easier. The Treaty is important as a symbol but it is also a legal text. Certainly the new Treaty provides a better frame for the EU as it moves into the twenty-first century. Without a strongly expressed environmental objective which recognises the needs of future generations, the EU is unlikely to hold the support of the European public, but ultimately continuing support depends on right action.

Drafted 1996 – published 1998

Developments since 1998

The Amsterdam Treaty, with the wording quoted above, was signed in June 1997. After being successfully ratified by all Member States it entered into force in 1999. When it was being signed a non-binding Declaration was attached to the Treaty committing the Commission to prepare 'environmental impact assessment studies when making proposals which may have significant environmental implications'. This was a strengthened version of a Declaration made at Maastricht and gave the Directorate-General for the Environment more power to insist that other Directorates-General integrated environmental considerations into their policies. The Treaty lead to the Sustainable Development Strategy and to the 'Cardiff Process' (see Chapter 1).

The Nice Treaty, which was signed in 2001 and entered into force in 2003, reformed the institutional structure of the EU to facilitate the significant enlargement of the EU in 2004. It made no changes to articles relating to the environment.

A proposed Treaty 'establishing a constitution for Europe' was signed in 2004 by the then 25 Member States but was then rejected in referendums in France and the Netherlands. After a period of reflection, when it must have been concluded that talk of a 'constitution' sounded too much like nation state building, the more modest Lisbon Treaty was signed in 2007 and, after being successfully ratified, entered into force in 2009. The Lisbon Treaty is a set of amendments to the two main Treaties governing the EU: the Maastricht Treaty (formally called the Treaty on European Union, TEU) and the Rome Treaty (formally called the Treaty establishing the European Community, TEC). The latter has been given a new name, the Treaty on the Functioning of the EU (TFEU). So there are now two main Treaties: TEU and TFEU.

Article 2 of the original Rome Treaty, discussed above at length, has been replaced by a much longer Article 3 of the TFEU. Section (3) of Article 3 now says that the EU is to work for sustainable development based on economic, social and environmental considerations – the three 'pillars' on which sustainable development is conventionally said to be supported. Brundtland's reference to the needs of future generations is acknowledged in the phrase 'solidarity between generations'. Article 3(3) reads as follows:

> The Union shall establish an internal market. It shall work for the sustainable development of Europe based on balanced economic growth and price stability, a highly competitive social market economy, aiming at full employment and social progress, and a high level of protection and improvement of the quality of the environment. It shall promote scientific and technological advance.

> It shall combat social exclusion and discrimination, and shall promote social justice and protection, equality between women and men, solidarity between generations and protection of the rights of the child.
>
> It shall promote economic, social and territorial cohesion, and solidarity among Member States.
>
> It shall respect its rich cultural and linguistic diversity, and shall ensure that Europe's cultural heritage is safeguarded and enhanced.

Article 3(5), which deals with relations with the wider world, says that the Union: 'shall contribute to peace, security, the sustainable development of the Earth'.

Note

1 Lord Avebury, President of the Conservation Society, in a letter published in *The Times* (London) 20 January 1975.

References

Brundtland Commission (1987) *Our Common Future*, Oxford: Oxford University Press.

Civic Trust (1974) Growth limits and the Treaty of Rome, *Civic Trust Newsletter*, January.

Commission of the European Communities (1984) *Bulletin of the EC*, 17(2), Luxemburg: Office for Official Publication of the EC.

ECJ (1983) Case 240/83 Procureur de la République v. Association de Défense des Bruleurs d'Huiles Usagées.

EEB (1979) *One Europe – One Environment – A Manifesto*, Brussels: European Environment Bureau, first published 1977.

EEB (1995) *Greening the Treaty II: Sustainable Development in a Democratic Union – Proposals for the 1996 Intergovernmental Conference*, Brussels: European Environment Bureau.

Haigh, N (1996) Sustainable development in the European Union Treaties, *International Environmental Affairs*, 8(1), pp 87–91.

Haigh, N (1998) Introducing the concept of sustainable development into the Treaties of the European Union. In: O'Riordan, T and Voisey, H, eds, *The Transition to Sustainability – The Politics of Agenda 21 in Europe*, London: Earthscan.

Haigh, N and Kraemer, A (1996) 'Sustainable development' in den Verträgen der Europäischen Union, *Zeitschrift für Umweltrecht*, 5(7), pp 239–242.

House of Lords Select Committee on the EC (1978) *Approximation of Laws under Article 100 of the EEC Treaty. Session 1977–1978*, 22nd Report HL 131, London: HMSO.

House of Lords Select Committee on the EC (1980) *Environmental Problems and the Treaty of Rome. Session 1979–1980*, HL 68, London: HMSO.

IEEP (1995) *The 1996 Inter-Governmental Conference: Integrating the Environment into other EU Policies*, London: Institute for European Environmental Policy.

O'Riordan, T and Voisey, H, eds. (1998) *The Transition to Sustainability – The Politics of Agenda 21 in Europe*, London: Earthscan.

von Moltke, K (1977) European Community: the legal base for environmental protection, *Environmental Policy and Law*, 3, pp 15–21.

Wilkinson, D (1992) Maastricht and the environment: the implications for the EC's environmental policy on the Treaty on European Union, *Journal of Environmental Law*, 4(2), pp 211–239.

4

AIR AND ACID RAIN

The Directive adopted in 1988 on emissions from 'large combustion plants' was an important milestone for the EU because it promised to resolve a long-running conflict between countries in Europe over acid rain. It was contentious not just because it entailed large costs, but also because the countries that would benefit most were not the ones that had to bear the largest costs. It subsequently acquired importance as a precedent for dealing with climate change as discussed in Chapters 2 and 9.

The following article was written for *International Environmental Affairs*, an American academic journal, after the Directive was agreed but before it was formally adopted (Haigh 1989). The evolution of the Directive is set in the context of pre-existing EU legislation and pollution control theory. The article demonstrates that the Directive was not just 'made in Brussels', contrary to a widely held view of how the EU operates, but was a truly collaborative venture, building on the efforts of a surprising number of international organisations and the suggestions of a number of different Member States over a period of years. No country on its own could have produced a solution and it was the EU, rather than other groupings of countries, that was able to negotiate an effective policy that resulted in action. The unique machinery of the EU provides the opportunity for almost continuous dialogue between officials and Ministers from different countries. An important impulse to produce a Directive had come from the German Presidency of the Council at a Summit meeting in Stuttgart in 1983. This noted 'the acute danger threatening the European forest area' and called for immediate action to avoid 'an irreversible situation'. It is said that German Chancellor Kohl would, over a period of years, ask the British Prime Minister Thatcher when the UK was going to agree to the Directive. Their terms of office (1982–88 and 1979–90, respectively) coincided with the exceptionally long negotiations (1983–88).

New tools for European Air Pollution Control – published January 1989

Until last June a question mark had been hanging over the environmental policy of the European Community (EC). Despite the absence of any mention of the environment in the founding Treaty of Rome, the EC has since 1972 been adopting a steady volume of environmental legislation that binds its 12 Member States. This has been providing an increasingly comprehensive framework which in turn is shaping national environmental policies so that it is now impossible to understand fully the environmental policies of many European countries without an understanding of EC environmental policy itself. While many of the 200 or more items of this EC legislation are of a technical character with little policy content, some are important by any standard. This is not the place to go into the reasons behind the EC's decision to embark on an environmental policy at all, but it is sufficient to say that they are bound up with the very purposes of the EC. They also include the two classic justifications for international collaboration in this field: the refusal of many environmental problems to be confined by national frontiers, and the impediments to international trade created by differing national standards. Leaving aside global issues such as the greenhouse effect, acid rain is arguably the most important environmental issue facing Europe today and one which clearly cannot be handled by individual European countries acting on their own. Any failure by the EC to deal with the acid rain issue would therefore call into question its ability to deal with the more difficult of the challenges for which it exists.

Although at the time of this writing the final text of the Directive on pollution from large combustion plants has not been adopted, its essential elements are known and were agreed in principle by the Council of Ministers in June 1988 (it was adopted as Directive 88/609). Apart from its importance in resolving a long-running dispute in European environmental policy, it is also important for its originality in giving a binding character to a new tool for pollution control. But it is certainly not the last word on the subject, and its very existence now poses with new urgency the question of the way forward for EC air pollution legislation. In attempting to answer this question, it is necessary first to review the existing tools for pollution control and the extent to which they have been used in EC legislation. In doing that, it helps to consider the theoretical underpinnings of air pollution control. These are usefully thrown into relief by contrasting two current concepts summarised by two contradictory statements. One is the famous dictum of Paracelsus that 'there are no poisons, only poisonous concentrations', and the other, a modern one, that states that 'dilution is no solution to pollution' (quoted in Simonis 1987).

The dictum of Paracelsus has stood the test of time and has the weight of scientific opinion behind it. Since so many toxic substances – many occurring naturally – are irretrievably dispersed, it would be alarming indeed if they were dangerous at even vanishingly low concentrations. We rely on the dictum as an operational concept when blending different sources of water supply to ensure that health-related concentrations of certain substances are not exceeded, and more prosaically whenever we open a window in a stuffy room.

The opposite statement that 'dilution is no solution' would, if completely true, rock society to its foundations. Nevertheless, the statement, although without scientific respectability unless heavily qualified, has enormous power and even some apparent common sense behind it. The British for example have known for some time that their air pollution policies of building power stations with tall stacks away from towns have enormously improved ground-level sulphur dioxide concentrations, by effectively diluting them. But they have also learned that it has done absolutely nothing to reduce their exports of sulphur dioxide and may even have increased them. The sulphur, much diluted, is being deposited somewhere.

The 'dilution is no solution' concept has, as a result of the acid rain phenomenon, acquired considerable force as a political slogan. But it nevertheless remains not much more than a slogan since it does not say what you should be doing instead. Logically it should mean you should emit nothing that is a potential pollutant when clearly this is often impossible. Probably it reduces to 'try to emit as little as you can' which in turn means 'use the best technology to prevent emissions'. A heroic attempt is being made in Germany to give greater precision to the slogan under the name of '*Vorsorgeprinzip*' – the principle of foresight (see Chapter 13).

The existing tools for pollution control

Let us bear these two powerful concepts in mind as we consider the existing tools for pollution control illustrated in Figure 4.1. This diagram is based on the classic

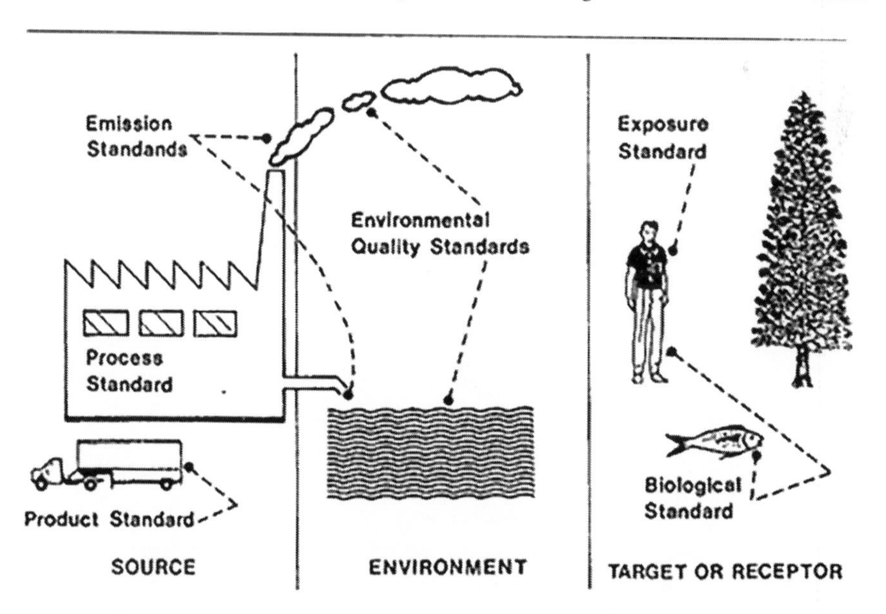

FIGURE 4.1 Points on the pollutant pathway at which standards can be set

definition of a pollutant as a substance causing damage to a receptor in the environment – a concept that is entirely consistent with the dictum of Paracelsus. Six points are shown along the pathway of the pollutant at which controls are traditionally applied. All of these controls had been used in national legislation or practice long before air pollution control became an international issue.

EC environmental legislation includes product standards in the form of limits on sulphur in fuel, and in the form of standards for vehicle emissions. It includes emission standards for factories discharging to air, and process standards, setting concentrations within factories to protect workers. It includes air quality standards and exposure standards, which in the case of air are really the same as air quality standards. And it includes biological standards such as acceptable blood lead levels, and standards for mercury in fish flesh.

Let us now consider approaches to air pollution policy with the help of the diagram and in terms of the two contradictory concepts that we started with.

If we are followers of Paracelsus, we logically enter the diagram from the right-hand side. If we can prevent 'poisonous concentrations' in the receptors rising above critical levels, all is well. In theory we can do this by setting standards for the quality of the environment which the receptors inhabit, and in turn we can ensure that these environmental quality standards are met by setting appropriate emission standards and product standards. Depending on where the sources of emissions are, and also the capacity of the environment to dilute or degrade the emissions, the emissions standards will vary from place to place.

Now the objection to this approach is that we are not omniscient. We cannot always identify in advance the critical receptor; we do not always know what is a critical concentration; and we do not always know the pathway by which pollutants reach receptors. The acid rain problem is a monument to our ignorance of these three matters.

If we are followers of the 'dilution is no solution' school, we logically enter the diagram from the left-hand side. Emission standards are then set as tightly as technology allows. The same goes for product standards for fuels and vehicles. Environmental quality standards and biological standards then just serve as a check to ensure that receptors are not being put at risk from non-point sources or from an excessive number of emitters.

In the field of water pollution control policy, there is a still unresolved doctrinal dispute within the EC. Some countries enter the diagram from the left with what can be termed a technological approach, and the British enter from the right with what can be termed an ecological approach (discussed further in Chapters 5 and 8).

In the field of air pollution there have been no basic differences between EC countries in pollution theory as represented by this diagram. Governments of all countries believe in a mixture of standards based on available technology for stationary plants (and for products), coupled with air quality standards to protect human health. None has legislation setting environmental quality standards to protect nature. In other words a mixed approach is adopted with the diagram being simultaneously entered from left- and right-hand sides.

These then are the traditional tools for air pollution control both in a national context and in EC policy. We can now look to see what new ones have been introduced to deal specifically with transboundary pollution.

Air pollution becomes an international issue

The following steps provide a brief summary of the rather complicated process by which air pollution became internationalised.

1. The work of Swedish scientists in the 1950s and 1960s first linked emissions and transport of sulphur from abroad with the acidification of Swedish lakes. A newspaper article by the scientist Svante Oden in 1967 prompted the Swedish government to limit the sulphur content of fuel oil in 1968.
2. In 1969, Oden's work was presented to the Organisation for Economic Co-operation and Development (OECD). In 1970 the OECD Council accepted a Nordic Council proposal for research work on long-range trans-boundary transport. This was begun in April 1972. (A first report was published in 1977 and a second in 1979, confirming that SO_2 [sulphur dioxide] moved long distances; Wetstone and Rosencranz 1983.)
3. At the 1972 United Nations (UN) Conference on the human environment at Stockholm, the Swedish Government was able to draw greater attention to its position as a 'victim' of air pollution from other countries.
4. In 1975 at the Helsinki Conference on Security and Cooperation, the Soviet Union proposed the environment as a possible subject for cooperation between East and West.
5. In 1977, as a result of the Russian initiative, and under pressure from Scandinavian countries, the UN Economic Commission for Europe (UNECE, which includes Eastern as well as Western European countries and, surprisingly, the United States and Canada) began work on an international convention on long-range transboundary air pollution. The Scandinavians pressed for a 'standstill' clause to prevent any increase in emissions and a 'roll-back' clause to reduce emissions by fixed percentages over the years to come.
6. In 1979 the Convention was signed at Geneva. As a result of resistance from the United States, Britain and West Germany, it contained no 'roll-back' clause but merely required signatories 'to endeavour to limit and, as far as possible, gradually reduce and prevent air pollution'.
7. In June 1982 at the Conference on Acidification of the Environment (held in Stockholm to mark the tenth anniversary of the great UN Conference), West Germany announced an about-face on its policy on air pollution as a result of growing concern about forest death in Germany. At the same time, West Germany submitted a memorandum to the EC Council of Ministers asking for priority to be given to a Directive on air pollution prevention.
8. In June 1983 at a meeting of the executive body of the Geneva Convention, the Nordic countries proposed a Protocol to the Convention requiring

signatories to cut SO_2 emission by 30 percent by 1993 (compared to a 1980 base). This was formalised at Helsinki in 1985. Countries that became parties to the Helsinki Protocol formed what was called the '30 Percent Club'.

9. In December 1983 the EC Commission, prompted by the German memorandum, proposed a Directive on emissions from large combustion plants. Five years later a Directive was adopted that contained both technology-based emissions limits and different overall percentage reductions for different Member States.

A new tool is born

From this brief survey, we can see the emergence of a new tool: the concept of a total national emission limit. This has been expressed more precisely as a percentage reduction by a future date over total emissions at some previous date. This first arose as a serious idea during the negotiations over the Geneva Convention in 1977 to 1979 in the form of the 'roll-back' clause that failed to gain acceptance. Following the German change of heart in 1982, the Scandinavians pressed it again and it was then formalised in the Helsinki Protocol of 1985. This tool is sometimes called a national 'bubble' by analogy with a rather different concept developed in the United States. In essence, an imaginary bubble is drawn around a given area (in our case, a country) and a limit put on the total amount of pollution from any source allowed to enter into that bubble. In the USA the concept has been used to encourage industrial plants in a given locality to trade their emissions among themselves so that the most efficient economic result is achieved.

Despite its shortcomings, the concept of a national bubble is attractive for its very simplicity. Some of the shortcomings of the national bubble concept involving a specific reduction by a specific date were emphasised in Britain when refusing to join the 30 Percent Club of parties to the Helsinki Protocol. Chief among the objections is that it is arbitrary. The base date of 1980 is arbitrary, and seems to have been chosen because it was the first year when data on emissions were available in most countries. The target date of 1993 is arbitrary, as is the percentage reduction. No-one seriously pretends that a 30 percent reduction can be related to the prevention of a given amount of damage or to a given improvement in the environment.

The advantages of the bubble concept as a tool in an international convention are that it involves no interference by one country in the internal policies of another. Indeed it reinforces the very idea of the nation state. Having accepted a given reduction, a country can achieve it by whatever means it likes. Examples would include industrial restructuring – which is sometimes a euphemism for industrial decline – or energy saving, adapting existing power stations, developing nuclear power or switching from coal to gas.

The first 30 Percent Club cannot be regarded as more than a crude first step. It is little more than a political commitment to move in the right direction. If the percentage reduction is not reached in a particular country, there is very little that other countries can do about it. An EC Directive embodying the bubble concept is a much sharper instrument, since the Commission has a duty to see the Member States fulfil their obligations and can refer the matter, if necessary, to the European Court of Justice. The EC Directive on large combustion plants is therefore original in giving the national bubble concept a binding character.

The EC Directive on large combustion plants

The Directive was proposed in December 1983 under pressure from West Germany, and its initial drafts were modelled on German legislation. For this reason these drafts included only technology-based emission limits. This is perhaps rather curious. It shows that the concept of 'national bubbles', developed as we have seen above under the UNECE Convention as the key tool to handle air pollution in an international context, had not yet gripped the minds of German officials and EC Commission officials responsible. This is even more surprising given that the bubble concept had been used in the EC in 1980 to limit total emission of chlorofluorocarbons as a precautionary measure to protect the ozone layer, as discussed in Chapter 2.

The bubble concept had, however, been introduced into the Directive by the time it was formally proposed. The 'bubble' part of the proposal took the form of national reductions of SO_2 (from large combustion plants only) of 60 percent (by 1995 from a 1980 base line) and NO_x [nitrogen oxides], and dust reductions of 40 percent. The same percentage reduction applied to each Member State. It was soon obvious that this would never be agreed and not just because of British objections, though Britain was the most consistent opponent of significant early reductions. The Commission made a tactical mistake in not recognising this earlier and as a result, the proposed Directive exerted no pressure on Britain in 1986 at a key moment when the Central Electricity Generating Board (CEGB) decided to retrofit three existing power stations totalling 6,000 megawatts with flue gas desulphurisation equipment (FDG). A more realistic proposal might have produced a greater response, as indeed the newly agreed Directive will do.

To shed some light on the difficult negotiations that finally led to agreement on the Directive, the Member States can be divided into four groups. The first consists of what can be termed the enthusiastic countries (West Germany, Netherlands, Denmark) which have a highly developed public opinion and are prepared to spend large sums on retrofitting existing plants. The second group consists of France and Belgium which are both effectively indifferent, since their existing nuclear programmes will ensure that the prescribed emission reductions will be achieved with no extra effort. The third group consists of Britain and Italy both of which are large emitters and dependent on coal, although Britain's high-sulphur domestic coal puts it in a more difficult position. The final group consists of the smaller countries, most of which are geographically remote from seats of damage caused by acid rain. Many of this last group are also industrialising rapidly and planning to increase energy consumption. The final text reflects these political and geographical realities and truly deserves its description as a compromise. It will nevertheless force many countries to move faster than they would have done otherwise.

In June 1986 the Dutch, holding the Presidency of the Council and accepting the fact that the Commission's original proposal would never be agreed, put forward a revised proposal that divided countries into groups with different reduction requirements by 1995, amounting to an overall EC reduction of 45 percent:

Group 1 (at least 50 percent) Germany, France, Belgium, Netherlands, Denmark
Group 2 (at least 40 percent) United Kingdom, Italy
Group 3 (standstill or 10 percent) Spain, Portugal, Greece, Ireland, Luxembourg.

The Commission and Germany thought this did not go far enough; others thought it went too far. However, the idea of different reductions for different countries was accepted. A feature of the Danish proposal put forward in December 1987 was for a three-phase reduction. For simplicity only the Danish figures for Germany, Britain and the EC as a whole are given in Table 4.1.

The stumbling block to this proposal was the third stage, which many countries felt was too far into the future for any figures to have any likelihood of being realistic. Nevertheless the idea of a three-stage reduction remained and is now embodied in the Directive that was agreed under the German presidency in June 1988. The reductions for SO_2 are given in Table 4.2.

The Commission's original proposal of a 60 percent reduction by 1995 has been effectively reduced by 3 percentage points and postponed by eight years. Instead of being achieved uniformly by each country, some countries will even be increasing their emissions.

A two-phase programme of reductions of NO_x was also agreed for different countries, amounting to an overall EC reduction of 13 percent by 1993 and 30 percent by 1998.

Since it is part of the Directive that all new power stations should be 'low-acid', that is, fitted with the best available technology for reducing emissions, the argument about percentage reductions by certain dates therefore amounts to an argument about how much money is to be spent replacing old stations or improving existing ones. For many countries the argument turns into how quickly they replace fossil fuel power stations with nuclear ones.

TABLE 4.1 SO_2 emissions reductions for the United Kingdom and the Federal Republic of Germany according to the December 1987 Danish proposal

	1993	*1998*	*2010*
United Kingdom	22	33	80
Federal Republic of Germany	40	60	80
EC	34	48	77

TABLE 4.2 SO_2 emission reductions by country according to the large combustion plant Directive

	1993	*1998*	*2003*
Belgium	–40	–60	–70
Denmark	–34	–56	–67
Germany	–40	–56	–67
Greece	+6	+6	+6
Spain	0	–24	–37
France	–40	–60	–70
Ireland	+25	+25	+25
Italy	–27	–39	–63
Luxembourg	–40	–50	–50
Netherlands	–40	–60	–70
Portugal	+102	+135	+79
Britain	–20	–40	–60
EC	–23	–42	–57

Possible new tool: critical loads

Most observers recognise that the 30 Percent Club was only a crude first step. The large combustion plant Directive is less crude, not only by involving different percentage reductions for different countries, but also by including technology-based emission limits for new plants. All new power stations will be 'low-acid' stations; that is, emitting as little acid as technically possible, irrespective of the commitments to a percentage reduction. Neither of these approaches is directly related to effects on the environment. We are still on the left-hand side of our diagram (Figure 4.1).

More refined approaches are under consideration in the academic community, most notably the RAINS model at the International Institute for Applied System Analysis in Austria (Hordijk 1986) and at the Beijer Institute, University of York. Both broadly follow the approach of building up a database on existing emissions and their geographical locations, the meteorological conditions which transport SO_2, and measured depositions of sulphur. There is then an attempt to assess the ability of different soils to absorb a given SO_2 deposition and then, working backwards, to assess the reductions that can be made at different SO_2 sources to ensure that the acceptable depositions are not exceeded.

Crucial to this approach is the definition of 'ecological target values' or 'critical loads' – the acceptable annual dose of sulphur that any particular area can take. We are right back to Paracelsus and the poisonous concentration. The 'critical load' is no more than another form of exposure standard or environmental quality standard. It is measured in grams per square metre per annum, as opposed to grams per cubic metre, in the case of air quality standards.

The great merit of this approach is that it is indeed an 'ecological' or 'environmental' approach rather than a 'technological' one. Its great difficulty is the difficulty we have already described. Even if one can be sure that one has derived a scientifically sound 'critical load', it is going to be extraordinarily difficult to apportion this back with any confidence to the sources in different countries that are contributing to the load. Such an apportionment would allow one to say with confidence that source A should be reduced by X percent, and source B by Y percent and so on. While conceptually it may be possible, in practice it will be extremely difficult to enshrine this in some workable convention or EC Directive.[1] The analogy with air quality standards which have been set with mandatory force in EC Directives is not a fair one since air quality is largely the result of local emissions and only rarely is affected by transfrontier pollution (at least at the rather high concentrations set in the EC Directives so far).[2] With critical loads this is not the case.

Experience with national administration suggests that the more complicated a piece of legislation, the more likelihood there is of it not being properly applied. This is even more likely to be true in an international context. The 30 Percent Club has the huge merit of simplicity. The difficulty of taking that one step further and apportioning different percentage reductions to different countries is illustrated in the difficulties over the large combustion plant Directive. It is therefore unlikely that the critical load concept will be enshrined in an international convention except possibly as a stimulus for countries to intensify their collective efforts and to provide a scientific justification for the expenditure of large sums of money. This is not to say that the effort should not be made.

If this sounds disappointing, it should be remembered that with the EC Directive agreed, then at least 12 countries are committed to having only 'low-acid' power stations in the long term. The task still remains to persuade other European countries to do the same. Here the critical load concept should play a powerful role.

Possible new tool: financial instruments

The concept of one country paying another to reduce emissions is known in the field of water. The Netherlands paid France not to discharge chlorides into the Rhine on the grounds that this would be cheaper than treating the water to make it suitable for drinking. The same argument has arisen with local air pollution and has been developed in the United States with the concept of emissions trading. This has hardly begun in an international context, although there may be opportunities here. It is cheaper for West Berlin to pay East Berlin to reduce their emissions than to continue to reduce their own emissions in West Berlin. The same could well be true of long-range transboundary air pollution, but we hardly have the international machinery for handling such a scheme. Possibly it could grow out of bilateral schemes.

Conclusion

The armoury of tools for air pollution legislation in a national context is primarily technology-based emission limits and product standards on the one hand, and environmental quality standards on the other. These will remain and be reinforced.

To deal with air pollution in an international context, an entirely new and simple tool has been introduced, namely the national bubble or total national emission limit with percentage reductions by a deadline. This will force the pace towards the adoption of the best technology. It is now being supplemented by the ecological concept of the critical load which is a form of quality standard. The difficulty of introducing this into legally binding form in an international context will be considerable.

Drafted November 1988

Developments since 1988

Of the three tools for air pollution control, described above as new in 1988, two have since been used in the EU for climate change. The EU-wide cap on emissions of greenhouse gases, with different Member States accepting a different share of the burden, was the cornerstone of its climate change policy from the beginning. The EU emissions trading scheme (EU-ETS) came a little later (see Chapter 9).

The third tool, critical loads, came to prominence when the Fifth Environmental Action Programme of 1993 stated that critical loads should not be exceeded across Europe. This led to the adoption of the 1997 Acidification Strategy (COM (97)88) that had been called for by Sweden when it joined the EU in 1995. It set out to reduce by at least 50 percent the area of sensitive ecosystems – estimated at 8.7 million hectares – exceeding critical loads by 2010. The strategy proposed many measures for reducing acidifying emissions – in addition to the 'large combustion plants' Directive 88/609 – of which the most important was the 'national emissions ceiling' Directive 2001/81. This was another example of the first tool: a cap on national emissions of certain acidifying substances from all sources with different ceilings for different Member States.

In addition to reducing acidification of water and soil, the 'national emissions ceiling' Directive 2001/81 is also intended to reduce ground-level ozone and eutrophication. It sets national emission ceilings for SO_2, NO_x, volatile organic compounds (VOCs) and ammonia (NH_3) that had to be met by 2010. The NO_x ceiling has proved the most problematic for the Member States. The European Environment Agency (EEA) reported in 2014 that SO_2 emissions in the EU had reduced by 74 percent by 2011 from a 1990 baseline. The most significant sources are energy production (58 percent), industry (20 percent), and commercial, institutional and households (15 percent).

The 'large combustion plants' Directive has since been amended and consolidated into the 'industrial emissions' Directive 2010/75.

There is an obvious link between vehicle emissions and fuel standards. Chapter 10 describes how the decision was taken in 1988/89 that small cars should be fitted with catalytic converters and how unleaded petrol became mandatory in the EU. By the mid-1990s the motor industry was complaining that the costs of more stringent vehicle standards were becoming excessive and that more attention should be paid to cleaner fuels. The result was the establishment of the European Auto-Oil Programme, which involved the Commission, the motor industry and the oil industry assessing the most cost-effective contributions from a range of measures to meet future air quality standards including vehicle standards, fuel standards, controls over evaporative emissions, and inspection and maintenance programmes. A number of Directives were subsequently adopted and a review of a second phase of the Programme (COM(2000)626) concluded that transport's share of overall emissions in 2010 would be significantly lower in 2010 than in 1990 for all pollutants except CO_2. The principal legislation on vehicle emissions is now the 'type approval' Regulation 715/2007.

The first Directive to set air quality standards was adopted in 1980 and related only to smoke and SO_2. Other standards followed, and in 1996 the 'air quality framework' Directive 96/62 provided for further standards to be set in daughter Directives. In 1998 the Commission launched an initiative called 'Clean Air for Europe' (CAFE) intended to avoid conflicts between measures proposed in different areas or at different times and to make it easier to select the most cost-effective measures. The subjects to be considered included air quality standards, emissions from stationary sources, fuel standards, vehicle emissions, and the acidification and ozone strategies. This could be seen as logical extension of the Auto-Oil Programme. The CAFE programme was published in 2001 (COM(2001)245), highlighting a number of issues and identifying particulates and tropospheric ozone as the most pressing problems but also stating that a range of other issues also needed attention, including nitrogen oxides, acid deposition and eutrophication.

The CAFE programme led to the 2005 'Thematic Strategy on Air Quality' (COM(2005)446) accompanied by a proposed revision of air quality standards that became the 'air quality framework' Directive 2008/50. The thematic strategy set out actions that it claimed would reduce the number of premature deaths in 2020 by 140,000 compared with 2000. It recognised that the 'emissions ceiling' Directive needed to be revised to set ceilings beyond 2010. The cost of implementing the Strategy was estimated at 7.1 billion euros per year, although health benefits were estimated at nearly six times as much, at 42 billion euros per year, while environmental benefits were not quantified.

Developing a revised 'emissions ceiling' Directive proved contentious. A round of analyses and meetings followed the 2005 strategy but it was not till December 2013 that a proposal (COM(2013)920) was finally made as part of the Commission's 'Clean air policy package' that also included a proposal to control emissions from medium combustion plants (COM(2013)919). The proposed revised 'emissions ceiling' Directive was withdrawn in 2014 by the Junker Commission, which has since said it will modify it to address the concerns that have been raised.

Several Member States have been found by the Court to be in breach of air quality standards, particularly the standards for NO_x, so EU air pollution policy remains contentious. The growing use of diesel cars has contributed to the problem. In the 1980s the main focus was on the acidification of lakes and forests. Today the great reductions in emissions that followed the adoption of the 'large combustion plants' Directive has resulted in Swedish lakes slowly beginning to recover. The effects of air pollution on nature are still an issue but the focus now is again on human health.

Notes

1 Critical loads were referred to in the 1988 Sofia Protocol on nitrogen dioxides to the 1979 Geneva Convention on long-range transboundary air pollution.

2 I would not have written this today. The higher quality standards now in place are frequently breached by pollution blowing across frontiers.

References

Haigh, N. (1989) New tools for European air pollution control, *International Environmental Affairs*, 1(1), pp 26–37.

Hordijk, L. (1986) Towards a targeted emission reduction in Europe, *Atmospheric Environment*, 20(10), pp 2053–2058.

Simonis, U. (1987) The Federal Republic of Germany. In: Enyedi, G., Gijswijt, A. J. and Rhode, B. eds, *Environmental Policies in East and West*, London: Taylor Graham.

Wetstone, G. and Rosencrantz, A. (1983) *Acid Rain in Europe and North America: National Responses to an International Problem*, Washington, DC: Environmental Law Institute, p 135.

5

WATER – TOWARDS CATCHMENT MANAGEMENT

In 1990 the *Financial Times* held a conference in London on the 'European Water Industry'. The timing was good as the EU had just proposed a Directive on 'urban waste water treatment' entailing very large costs, and there was much discussion of privatising public services, including drinking water and sewage disposal. The speakers and participants came from around Europe and I was invited to give 'an environmental view'. Rather than recite a catalogue of the concerns of environmentalists, I chose to ask where water then stood in the hierarchy of environmental problems, before considering what a number of international organisations had been saying about water (Haigh 1990).

At that time there were only twelve EU Member States, of which a majority are largely self-contained for water. Some are peninsulas (Italy, Greece, Denmark, with Spain and Portugal together forming the Iberian peninsula), and some are islands (United Kingdom and Ireland). Even France, whose frontier with Germany is partly defined by the Rhine, is otherwise largely self-contained, since the other five sides of the great French hexagon are defined by the Alps, the Pyrenees, and seas. Of the original six Member States, all except Italy have territory within the catchment of the Rhine, with the Netherlands being almost entirely dependent on it for water. The problems of the Rhine, a river that drains one of the world's most heavily industrialised areas while at the same time providing drinking water for a large population, thus dominated EU policy-making, with the Netherlands being particularly engaged. This was despite the enlargement of the EU in 1973 to include the UK, Ireland and Denmark, and later Greece, Spain and Portugal. In 1990 Portugal was the only other Member State that heavily depended on a neighbour for its water.

With its subsequent eastward enlargement, the EU now contains many more transboundary rivers, with the Danube being the most international river in the world as its catchment extends into no fewer than nineteen countries. It flows directly through or along six EU Member States (Germany, Austria, Slovakia, Hungary, Romania and Bulgaria) on its way to the Black Sea.

In my talk I noted how the dominance of the Rhine began to change at a Ministerial seminar held in Frankfurt in 1988 on the future of EU water policy. This led to the 'urban waste water treatment' Directive of 1991 and to the 'water framework' Directive of 2000 that was to take a much more comprehensive view of water management than before.

Water quality – an environmental view – talk delivered in London, March 1990

According to the Greek philosopher Empedocles, all matter was composed of four elements – earth, air, fire and water – and the idea persisted well after the Middle Ages.

One can see the power of the idea. Leaving aside fire, the other three are the elements of the environment that surround us and that support all life.

They are also the sectors into which environmental policy is today divided for the convenience of legislators and administrators.

As I am asked to talk about the environment at a conference dealing with only one of its elements, the first point to make is that the environment is ultimately indivisible.

The administrative demarcation lines with which we are familiar are merely a convenience. They can also become an impediment to the adoption of the best solutions to our problems, and the emerging emphasis on integrated pollution control is a welcome recognition of this fact. We have learnt that major inputs of metals to the North Sea are deposited from the air. We know that landfill sites and the application to land of fertilisers and pesticides are sources of water pollution. We also know that reducing emissions to one element can simply result in shifting them to another. Our three elements do not exist in separate compartments.

But since the conference has accepted the division, let me stick with it and ask where water stands in the hierarchy of environmental problems today?

If one takes the three examples of environmental problems – the greenhouse effect, the depletion of the ozone layer, and acid rain – singled out for special attention by Mrs Thatcher at her famous speech to the Royal Society eighteen months ago, only acid rain is obviously connected with water. But here too its solution falls largely in the field of air pollution policy.

I think it fair to say that water pollution is not seen as a global issue in the same way as the greenhouse effect, and the ozone layer. Nor is it seen as such a wide ranging regional issue as acid rain. All three are seen as aspects of air pollution. If the sea level rises, that will not be the fault of the European Water Industry as I understand matters.

There are good physical reasons why water is not a global issue. Our three environmental media through which pollutants travel – air, water, soil – are embodiments of the three physical states – gas, liquid and solid. As gases are lighter than liquids, so under the influence of gravity water stays on or near the Earth's surface. Gravity ensures that rain falls and then follows more or less prescribed channels into the seas. Water, unlike air, does not blow where it wills. Though water does not stay tidily where it is put, its movement is relatively circumscribed.

We are not yet at the point where pollution of the North Sea is a threat to the Indian Ocean.[1]

Does this mean that water problems are purely local? The map below will immediately show that this is not the case. Here we see the catchment areas into which the rivers and seas of Europe are divided.

There are seven seas which are the ultimate receptors of all European rain. The Baltic connects with the North Sea which connects with the Atlantic. The Black Sea flows into the Mediterranean which is also fed by the Atlantic. There is also the Arctic Ocean and the Caspian Sea which is an entirely inland sea. Of the rivers that feed these seas many cross national boundaries.

Clearly the protection of seas is not just a local nor even just a national problem. It must be handled internationally, as must the protection of many rivers. These catchments shown on the map do not neatly coincide with international groupings of countries and none of the seas, for instance, is abutted only by EC Member States. As a result there are many separate international conventions dealing with protection of individual seas and individual rivers.

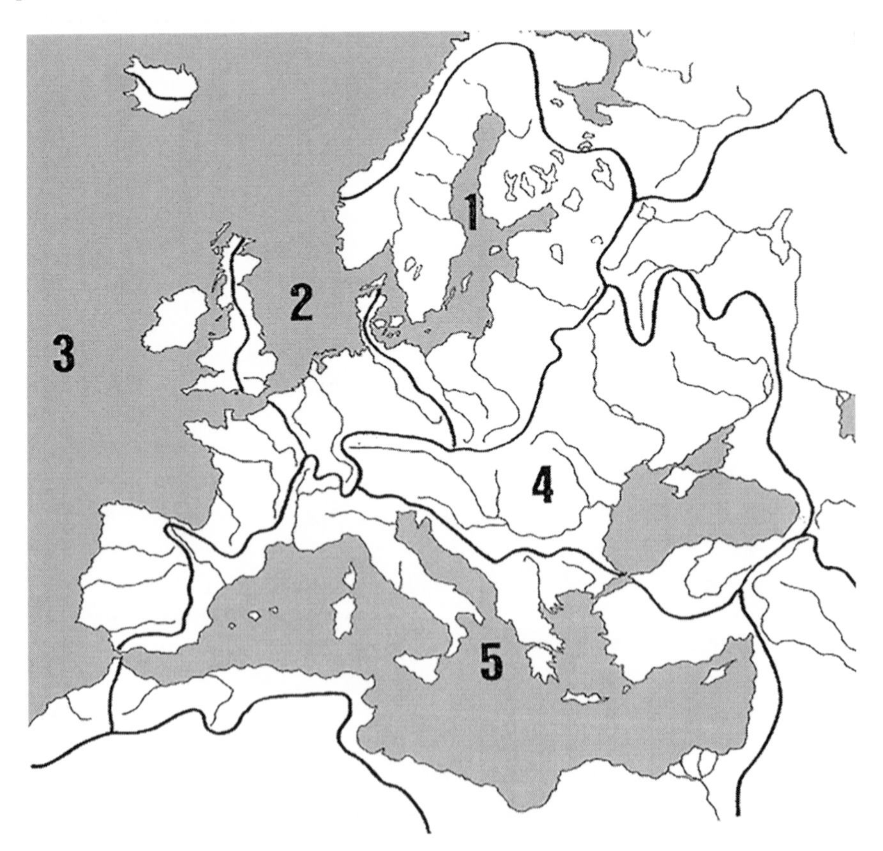

FIGURE 5.1 European catchment areas: 1 Baltic Sea, 2 North Sea, 3 Atlantic Ocean, 4 Black Sea, 5 Mediterranean Sea

As the geographical variation in Europe is wide, so are environmental problems. Let me just take two examples.

The Black Sea has an abiotic bottom layer poisoned by hydrogen sulphide and so has been suggested for use as a permanent waste-disposal site. It is entirely natural and was thought to be stable. Since it is already poisoned, why not concentrate poisons there, went the argument. The top of the layer is gradually rising and some believe that one day it will breach the surface with consequences that are unknown, although some say there are good scientific reasons for that being impossible.

My second example is the river Loire in France, which is one of the few great European rivers that has not been tamed with weirs and locks. A proposal to dam the upper reaches to control floods – and to provide a steadier supply of cooling water for nuclear power stations – has recently been postponed while alternatives are considered.

While the Black Sea affects several countries – not all of them in Europe – the Loire is an internal matter for the French although nature conservationists from all over Europe are keenly watching the outcome.

These are just two of the many problems of the water environment in Europe. To bring some order to what could otherwise be a catalogue of environmental issues, let me consider how they have been thought about by a number of international organisations or groupings.

The OECD (Organisation for European Co-operation and Development) last year adopted a Council Recommendation dealing with three subjects:

- integrated management of water resources and other policies (e.g. agriculture, industry, forestry)
- water demand management
- protection of groundwater.

The preamble to the Recommendation gives a flavour of the reasons. It recognises that water is in limited supply in many areas, and at certain times, and that the cost of meeting rising demands for all water services of the appropriate quality, in an environmentally acceptable way, is increasing rapidly. It recognises that the alternative to responding to rising demand through increasingly expensive provision of new supply, which could have significant effects on the environment, is effective demand management. It recognises that groundwater is a high-value scarce resource increasingly threatened by pollution.

An OECD recommendation has no binding force, but it requires governments at least to think about the points raised. One can perhaps see the concerns of the southern – and drier – European countries being expressed here in the emphasis on demand management. The groundwater issue will certainly be forced more into public attention in the Member States of the EC by the groundwater Directive of 1980 that is now being seen to have major implications for waste disposal sites.

The European Community (EC) began its environmental policy in the mid-1970s with a series of Directives dealing with water quality for certain uses: bathing, abstracting drinking water, fish life and drinking water itself. It simultaneously embarked on setting numerical emission standards for dangerous substances coming from point sources, mainly industrial plants, based on best available technology.

The bathing and drinking water Directives are well known. They introduced mandatory numerical standards for the first time and are causing painful costs in several countries. They are also focusing attention on a number of issues, notably nitrates and pesticides. The water for fish life Directives have been largely a dead letter since they appear to give the Member States discretion to designate no waters to which the standards apply. The German Länder for example have said they are not going to designate any. A recent judgement of the European Court has given the Commission confidence that they may be able to compel some designations and we can await events.

The dangerous substances Directive was driven by the problems of the Rhine and has concentrated attention on applying the best available technology to minimising discharges of dangerous substances. It forced the UK eventually to abandon its earlier insistence on setting emission standards solely by reference to quality standards in the receiving water. The realisation that substances must be accumulating somewhere, even if locally environmental quality was being preserved, provided the intellectual argument for the change of policy – this is an example of the precautionary approach being followed. But the paradox is that since most of the substances are pesticides anyway and are being emitted from rather few point sources, the policy is not going to work without environmental quality standards too to provide a reference point for the diffuse sources.

The domination of EC policy by problems of the Rhine, and by dangerous substances from point sources, began to change at a Ministerial seminar on the future of EC water policy held in Frankfurt in June 1988. Other environmental problems instead began to come to the fore. The conclusions of the seminar included the following points.

1. The need to *integrate problems of water quality with water quantity* (till now there is no EC policy on water quantity).
2. The need for EC legislation covering *ecological quality of surface water.* A draft Directive is now being worked on with the objective of maintaining the capacity of water for self-purification, preserving the diversity of species, and protecting the quality of sediments.
3. The need for adequate treatment of *sewage.* The point here is that many European towns have no sewage treatment – some of them inland cities such as Brussels and Milan. The Commission has now proposed a Directive, which also sets standards for nutrients (phosphorus and nitrogen) discharged by sewage works to 'sensitive areas' to be defined by the Member States.
4. The need to reduce pollution from *diffuse sources*, in particular nutrients and pesticides from agriculture. There is now a proposed Directive to limit the use of nitrogenous fertilisers on agricultural land.
5. The need to deal with *dangerous substances* simultaneously by the quality objective and emission standard approaches.

The EC Frankfurt seminar shows a shift away from a concentration on technological solutions applied to industries to a much broader approach that recognises the need to measure desired environmental quality against certain criteria (the forthcoming Directive on ecological quality of surface water) and that focuses on the major non-industrial sources of water pollution, namely sewage and agriculture.

Instead of being driven by the problems of the Rhine, EC water policy looks as if it is now being driven by eutrophication and the lack of investment in sewage works. The eutrophication problem is at least in part the consequence of over-intensive agriculture which in turn is driven by the EC Common Agricultural Policy. The lack of investment in sewage works is a particular problem in Southern Europe.

One can also say that EC policy is being driven by itself because of the need to meet the standards for nitrates and pesticides in the drinking water Directive. I am already on record as saying that the drinking water Directive will fall into disrespect if some of it is not revised.

The recent North Sea Conference held at the Hague is the third in a series, and a fourth is promised. The final declaration is an astonishing catalogue of problems and promised solutions, and I cannot mention them all. It talks about:

- control and enforcement of regulations
- developing low-waste technologies
- the application of the precautionary principle which it restricts to toxic and persistent substances
- an integrated approach considering emissions to both water and air
- the combined use of environmental quality objectives and emission standards
- reducing the risk of accidents.

It then proposes action:

- to achieve percentage reductions in dangerous substances
- to phase out and destroy PCBs [polychlorinated biphenyls]
- to reduce nutrient inputs involving sewage works, industry and agriculture
- ending the dumping of sewage sludge and sea incineration
- enhancing scientific knowledge
- protecting habitats and species.

The programme is comprehensive and ambitious. It recognises that the environment cannot be divided into three separate elements and hence the need for an integrated approach. The fact that Switzerland was represented is a recognition that all countries in a catchment have a responsibility and not just countries bordering the North Sea.

Some of the promises are precise while others remain merely vague but pious hopes. Some will cost a lot of money. It is increasingly difficult for environmentalists to complain that their concerns have been overlooked – all they can do is say not enough is being done. And of course they do. It is indeed not clear whether the expectations raised will be realised.

One observation I make is that there is no reference to the use of taxes or other economic instruments as a way of achieving the desired results. A subject that is so fashionable in the air pollution field does not yet seem to be high on the agenda of those concerned with water in an international context.

I come finally to the Council of Europe whose Committee of Ministers in 1968 proclaimed a European Water Charter setting out twelve principles. It is still worth reading today partly because it provides a measure of how far we have travelled in 22 years. It does not talk about technology at all, let alone best available technology, nor about industry, nor waste minimisation, nor the precautionary principle, nor integrated pollution control. On the other hand, it does talk about the need for river basin management, for international cooperation, for maintaining the quality of water at levels suitable for the use to be made of it, and for the need for assessment of water resources and their quality.

This last point cannot be over-emphasised. Let us assume that the 50 per cent and 70 per cent reductions in emissions promised by the North Sea Conference are fulfilled. How are we to know whether that is enough? Only if we have reliable assessments of the quality of our rivers and seas, and standards of quality by which to measure our efforts to reduce emissions, will we know what to strive at or whether we have arrived. This is particularly true of diffuse sources such as pesticides. We have yet to establish acceptable loadings.

The British have a particular role to play here. The British insisted in the 1970s that the right way forward was the setting of environmental quality standards, and refused to accept that emissions for toxic and persistent substances should be set by relying solely on best available technology. Having made a dramatic change of policy in 1987, Britain is well placed to insist again on the quality objective approach. Both the EC and the North Sea Conference have recognised the need for that approach but it is not clear what emphasis is being given to it. Will the Germans only implement the freshwater fish Directive if they find themselves in the European Court? Why do they not do so now as part of the Rhine plan? The issue is very much a live one.[2]

Perhaps the quality objective approach will be given a boost by the Council of Europe's latest initiative. A few days ago it launched its 'Save the Fish' campaign on the initiative of the Swiss Nature Protection League. This is concerned with the diversity of freshwater fish species and the threats to them and their habitats. The campaign propaganda makes the alarming claim that of the 200 European freshwater fish species, 103 are in immediate danger.

Eleven countries are participating in the campaign including for the first time an Eastern European country – Czechoslovakia. Why is Britain not among them?

These countries have promised practical action to inform the public, schools, specialists, decision-makers, farmers and industry about fish diversity and the environmental needs of fish.

When we know that all the species are not declining but are flourishing wherever they should, we will have a good indicator that we have arrived at our goal.

Responsibility for our indivisible environment rests at least partly with Europe's water industry.

Conversely the measure of the success of the water industry resides in the health of the environment. I hope all of you in the European Water Industry are ready to rise to the challenge.

March 1990

Developments since 1990

Two of the ideas floated at the Frankfurt seminar of 1988 have since transformed EU water policy: the need for adequate sewage treatment, and the need to focus on the ecological quality of surface water. Providing adequate sewage treatment was so expensive for the poorer Member States that a special fund – the Cohesion Fund – had to be created when the Maastricht Treaty was being negotiated in 1990. This, like acid rain and the other topics mentioned in Chapter 1, again raised the environment to the level of high politics.

The 'urban waste water treatment' Directive 91/271 was proposed eighteen months after the Frankfurt seminar. It required all towns and villages of over 2,000 inhabitants – or more accurately 'population equivalents', since animal waste also enters sewers – to have sewerage systems, with larger towns requiring more stringent treatment. Higher standards were required for discharges to particularly sensitive areas. The dumping of sewage sludge at sea was banned and its disposal on land controlled. The Directive would obviously involve very considerable costs, and by chance the Maastricht Treaty was being negotiated at the same time as the Directive. Accordingly Spain was able to insist on a special fund of 15 billion ecus (equivalent of euros) – known as the 'Cohesion Fund' – to be available until 1999 for those Member States with a GDP less than 90 per cent of the EU average (Portugal, Ireland, Greece and Spain). The fund was to be spent on transport and environmental projects including sewage works. Thus a specific item of environmental legislation triggered a new financial transfer mechanism from the richer to the poorer Member States. It could be said that the fund was the price extracted by Spain for agreeing to monetary union and thus to the adoption of the euro. The fund was later extended to include other countries that subsequently joined the EU.

The 'urban waste water treatment' Directive 91/271 was soon supplemented by the 'nitrates from agricultural sources' Directive 91/676, also to prevent eutrophication and to help meet standards for nitrates in the 'drinking water' Directive.

Despite delays in implementing Directive 91/271 in many Member States, and despite frequent infringement proceedings in the Court of Justice, many sewage works have been built or upgraded, and the European Environment Agency has been able to report considerable improvement in water quality.

The proposal for a Directive on the 'ecological quality of surface water', also called for at Frankfurt, took much longer to evolve because of its complication and range. What was eventually adopted in 2000 as the 'water framework' Directive 2000/60 radically changed EU water policy. It requires Member States to publish river basin management plans – which may require collaboration between Member States[3] – identifying all human pressures and containing information on the measures to be adopted to achieve 'good' water status for all surface water (lakes, rivers and coastal waters) and groundwater. All waters have to be classified into different categories of ecological status according to their biological, chemical and hydro-morphological characteristics. The plans are to be updated periodically. Because of the challenging requirements of the Directive, many Member States have shown little appetite to adopt the measures needed. Greater emphasis is now being given to improving the second river basin management plans to be adopted at the end of 2015.

Two further Directives have reinforced the catchment, or river basin, management approach. Although the 'water framework' Directive referred to flood management as an issue, it was imprecise about what was to be done. The 'flood risks' Directive 2007/60 requires an assessment of the flood risk of each river basin and their associated coastal zones, and flood risk management plans, in collaboration with neighbouring countries where necessary. The 'marine strategy framework' Directive 2008/56 applies to all marine waters and requires plans to be drawn up for certain 'Marine Regions': Baltic, Mediterranean, and Black Seas, and the North-East Atlantic Ocean (which includes the North Sea). It lacks the rigour of the 'water framework' Directive, and while it provides a new approach to understanding the problems facing Europe's seas, it is yet to be seen if it will lead to action to solve those problems. Coordination between Member States is encouraged, including by working through existing regional seas conventions.

One other Directive must be mentioned as it lays to rest the long-running dispute about the relative merit of environmental quality standards and emission standards for dangerous substances. The origins of the dispute in the 1970s, and the compromise that allowed the Member States a choice between the two approaches so that neither was obligatory, has been described at length elsewhere (Haigh 1984). Directive 2008/105 on 'quality standards for water' now makes such standards mandatory for surface waters (or exceptionally sediments and/or biota) and sets them for thirty-three 'priority substances' and eight other pollutants in order to achieve the objective of 'good status' required by the 'water framework' Directive 2000/60. Discharges of thirteen 'priority hazardous substances' – ones that are toxic, persistent and liable to bioaccumulate – are to be phased out within twenty years. Where a Member State cannot meet a quality standard because of transboundary effects, it will not be in breach of the Directive provided transboundary coordination mechanisms are in place. Emission standards are not set in Directive 2008/105, but these are set for individual plants under the 'industrial emissions'

Directive 2010/75 (see Chapter 8). Thus the call made at Frankfurt for the combined use of both approaches has been fulfilled.

Notes

1 This was my understanding at the time. Subsequently the 2002 Stockholm Convention on Persistent Organic Pollutants (POPs) sought to eliminate certain chemicals because they can be transported long distances not just by air but also by water. It is now also well known that floating plastic material is transported long distances by ocean currents. My argument still holds for pollutants that degrade fairly quickly.

2 Martin Gruner, Parliamentary State Secretary, German Ministry of the Environment was a speaker at the conference. We corresponded after it and he wrote to me (18.5.90) to say 'The recent ECJ judgement you mention ... will naturally be respected in the FRG. Consequently, implementation of the Directive is now being started in the FRG, too.'

3 For example the International Commission for the Protection of the Danube River (Sophia Convention) now acts as the platform for the implementation of all transboundary aspects of the EU 'water framework' Directive.

References

Haigh, N (1984) *EEC Environmental Policy and Britain – An Essay and a Handbook*, London: Environmental Data Services.

Haigh, N (1990) European water quality – an environmental view. In: *Financial Times Conferences: The European Water Industry*, conference proceedings, 26–27 March 1990.

6

FROM WASTE TO RESOURCES

Waste is usually thought of as being solid – something that can be transported in lorries and dumped on land – though waste materials are also emitted in liquid or gaseous form. Successful efforts to reduce industrial pollution of air or water often result in the generation of more solid waste, so in some senses pollution of air, water and land is interchangeable (see Chapter 8). What is blown or flows across frontiers is an obvious subject for EU environmental policy, but solid waste moves only when it is to be disposed of or recycled. Although the definition of waste has generated much debate within the EU, a common-sense view is that it is material of no value which someone has discarded or wants to discard.[1] Anyone generating waste, be it hazardous or not, will accordingly wish to be rid of it in the cheapest and most convenient way within the constraints of whatever rules apply. One possibility is to transport it across a national frontier to a country with laxer rules, or no rules at all. This was one justification for common standards for waste management being set at EU level, but that still left open the possibility of transport to countries outside the EU. Indeed, much EU waste was once exported to third world countries.

For a long time there were two schools of thought about the transfrontier shipment of waste. One held that it was a legitimate international trade if properly managed. This is consistent with the provision for the free movement of goods in the EU Treaty, on the assumption that waste was 'goods', a view that was disputed until the Court ingeniously resolved the matter in 1992 (see below). The other point of view was that waste is inherently different from valuable goods since anyone handling it has every interest in getting rid of it. To turn a blind eye to where it goes, or deliberately to dispose of it illegally, is a constant temptation, and it is no surprise to find that criminal elements have been involved in waste management in many countries, if management is the right word in those circumstances.

Although waste was among the first topics to be covered by EU environmental legislation, it was not till 1989 that the Commission, in its 'Community Strategy for Waste Management', discussed the desirability of restricting its movement. The need to prevent the generation of waste, to encourage recycling, and so to save materials and reduce the volume for disposal had been recognised in both the first 'waste framework' Directive 75/442 and the first 'dangerous waste' Directive 78/319, but expressed only as aspirations. The strategy of 1989 was written in anticipation of the completion of the internal market – promised for the end of 1992 by the Single European Act – and so warned that 'in a Community without internal frontiers the flow of waste towards lower-cost disposal plants may become a flood'. The strategy noted that there was a lack of sufficient, technically viable and suitably located disposal plants within the EU, particularly incinerators, and that a large quantity of waste was exported each year outside the EU. Harmonising high disposal standards was said to remain a priority, but the Commission also saw the need to reduce movement by favouring disposal 'in the nearest suitable centres'. This was named the 'proximity principle' in the second EU waste strategy of 1996 but remained as something desirable rather than mandatory. The 'waste framework' Directive 75/442 was nevertheless amended shortly after the first strategy to place a duty on Member States to establish an adequate network of disposal installations to enable the EU to be self-sufficient in waste disposal.

For obvious reasons the debate on controlling movement first concentrated on hazardous waste. The 'dangerous waste' Directive 78/319 made no attempt to restrict transport, confining itself largely to defining what was hazardous, and requiring those transporting and handling it to keep records and to be subject to authorisations specifying methods of disposal and the precautions to be taken. Following a very public scare, when drums thought to contain soil contaminated with dioxin from Seveso in Italy (where a factory had exploded in 1976) were lost in transit and were later found in France, the 'transfrontier shipment of hazardous waste' Directive 84/631 was adopted requiring anyone moving hazardous waste across frontiers, both within and out of the EU, to inform the authorities in the receiving country. In other words, international trade in waste was still thought legitimate if properly managed. The United Nations Environment Programme (UNEP) then recommended that exports to third world countries should take place only when they had the necessary disposal facilities, and the Organisation for Economic Co-operation and Development (OECD) in 1985 recommended that OECD countries should not allow exports without the consent of the receiving countries. This led to the 1989 Basel Convention on the 'control of transboundary movements of hazardous waste and their disposal', which enabled parties to ban the import of waste. Directive 84/631 was accordingly replaced by Regulation 259/93 on the 'supervision and control of shipments of waste within, into and out of the EC'. Among its many provisions it banned the export from the EU of waste for disposal (except to EFTA[2] states which are parties to the Basel Convention).

It was a coincidence that in the year of the Basel Convention a geopolitical event of the greatest importance, quite unconnected with the environment, was to give the 'proximity principle' and national self-sufficiency much greater prominence. This was the fall of the Berlin wall in 1989 which led to the reunification of Germany one year later.

For many years before German reunification, a huge landfill site at Schoenberg in the German Democratic Republic (GDR), just over the border with West Germany and not far from Lübeck, had been receiving both hazardous waste and domestic waste from the GDR and from some other communist bloc countries. But it also received waste from some EU Member States. This suited everyone. The GDR was glad to earn hard currency, and nearby EU Member States, including West Germany and the Netherlands, had an accessible outlet that relieved the pressure on sites within their own territory, where for geological reasons safe landfill is expensive. It was thus possible for West Germany to maintain high disposal standards on its own territory while at the same time making use of unmanaged and unregulated sites just across its border. When Germany became reunified in 1990, West German legislation became applicable in the former GDR, so the Schoenberg site and others like it had to be closed almost immediately. This had obvious repercussions for German waste policy as its existing sites were probably soon to be overloaded by waste from the former GDR as well as surplus West German waste that previously went to Schoenberg. German waste policy was accordingly forced into placing much greater emphasis on waste reduction and incineration, though it is only fair to say that German policy was already moving in that direction. One result was the German packaging ordinance, which obliged producers of packaged goods to take back the packaging or pay someone else to do so. The possibility of waste moving to new outlets nevertheless created pressure on neighbouring countries. All these German developments were to influence EU policy, and the question of whether waste was 'goods' with the right to move freely now needed to be resolved urgently. Fortunately the 'Walloon waste case' provided the occasion.

The Walloon Region of Belgium had issued a decree in 1985 placing an absolute prohibition on the storage and disposal of non-Walloon waste whether hazardous or not, and in 1990 the Commission began proceedings against Belgium before the Court on the grounds that the Walloon decree breached waste Directives as well as the Treaty provision on the free movement of goods. In its judgement of July 1992 (Case C-2/90), the Court held that wastes were indeed 'goods' irrespective of whether they were to be reused, recycled or disposed of, but then went on to say that wastes are goods of a specific nature which may constitute a risk to the environment so that their free movement may be limited for reasons of environmental protection in accordance with existing legislation. The Court held that Belgium had not breached the 'waste framework' Directive 75/442 which said nothing about movement of waste between Member States, but did breach the 'transfrontier shipment of hazardous waste' Directive 84/631 which said – in a recital – that transfers of waste may be necessary between Member States in order to dispose of it under the best possible conditions. The Court's decision effectively allowed Member States to make the 'proximity principle' mandatory, and one month later France issued a decree banning the import of domestic waste intended for landfill.[3]

It was under these influences that in 1996 the Commission issued its second waste strategy, which it called a review of the first. It marks a shift in EU policy away from just waste management to resource recovery[4]. The word 'recovery' was used to mean recovery of materials and energy, and so is wider than 'recycling', which is usually restricted to materials. Where there was a choice, the new strategy preferred the recovery of materials to recovery of energy, in other words it preferred recycling over incineration (since energy recovery means incinerating waste and using the heat generated for heating or generating electricity). The new strategy reaffirmed the waste 'hierarchy' of the first strategy (in descending order of preference: prevention, recovery, safe disposal – the hierarchy applying to Member States and not to waste operators). It reaffirmed, and named, the principles of proximity and self-sufficiency in waste disposal facilities, and gave currency to the phrase 'producer responsibility'. While recognising that many shared the responsibility for waste during the life cycle of a product, it was the product manufacturer who had the predominant role 'being able to conceive products in a way which facilitates proper reuse and recovery'.

The Council, in responding to the first strategy, had asked for action on particular types of waste and the Commission had accordingly instituted a 'priority waste streams programme'. This focused on tyres, end-of-life vehicles, healthcare waste, construction and demolition waste, and waste from electrical and electronic equipment. Several items of legislation were then adopted embodying 'producer responsibility': batteries (1991), packaging (1994), end-of-life vehicles (2000), and electrical and electronic equipment (2002). Each of these required the products to be designed so they were easy to recycle, restricted the use of dangerous substances that made recycling difficult and, in some cases, placed a duty on the manufacturer to take back the product at the end of its life, or arrange for this to be done. The 'packaging' Directive 94/62, inspired by both the German packaging ordinance and a draft French law, was intended not primarily to reduce the amount of packaging on products, but to limit the amount of packaging waste going to disposal. Like the 'landfill' Directive 1999/31 (see below), it has required Member States to arrange separate collection of different wastes. In 1992 the 'ecolabel' Regulation 880/92 was adopted to give consumers greater information about products with reduced environmental impact, thus applying pressure on producers to market them. (The much later 'ecodesign' Directive 2005/32, although concerned with energy efficiency standards for products, provides a precedent for setting standards for recyclability and recycled content.)

Following the first strategy, the Commission had proposed a Directive setting standards for landfill, but after lengthy discussions the Parliament effectively killed it in 1996 as not being stringent enough. The Commission continued to receive complaints about poor standards, and in September 1996 100,000 tonnes of waste from a landfill near la Coruña in Spain slipped down a hillside endangering the nearby town and threatening to spill into the sea. In addition, a wholly new argument emerged when it was estimated that 32 per cent of the overall release in the EU of methane, a powerful greenhouse gas, arises from decomposing waste in landfills. Very little of this methane was being captured for energy or even burnt off. A new proposal was therefore put forward requiring phased reductions in the overall amount of biodegradable municipal waste going to landfill. After much dispute the 'landfill' Directive 1999/31 was adopted with the main aim of reducing emissions of methane, but also encouraging prevention, recycling and recovery, and reducing shipment.[5] The Directive prohibited the landfilling of certain wastes, including liquids and tyres. Landfills were divided into three classes – hazardous waste, non-hazardous waste and inert waste – with a prohibition on these wastes being mixed. By setting deadlines for reducing biodegradable waste in landfills, the Directive has forced Member States to arrange for separate collection of such waste.

The Sixth Environmental Action Programme of 2002 had promised a number of 'thematic strategies' including one on waste and one on resource use, though obviously the two remain intertwined. The 'Thematic Strategy on Waste Prevention and Recycling' (COM(2005)666) (which can be thought of as the third waste strategy, following the first of 1989 and the second of 1996) set some underlying principles, including taking a life-cycle approach to policy-making, a new focus on waste prevention, and a shift to a materials-based approach instead of focusing on particular types of end product such as under the priority waste streams programme. A number of key policy changes were also envisaged. In 2011 the Commission published a report claiming that progress had been achieved in improvement and simplification of legislation; the diffusion of concepts such as the waste hierarchy and life-cycle thinking; increased focus on waste prevention; and setting new collection and recycling targets. Recycling rates had improved, the amount of waste going to landfill had decreased, the use of hazardous substances in some waste streams had been reduced, and the relative environmental impacts per tonne of waste treated had decreased. These achievements, however, were offset by the negative environmental impacts caused by increased waste generation.

One example of the improvement and simplification of legislation was the 'waste framework' Directive 2008/98, which replaced and brought together three Directives: the previous 'waste framework' Directive; the 'hazardous waste' Directive; and the 'waste oils' Directive. It did much more than consolidate these by introducing into legislation many of the concepts developed in the preceding strategies. It retained some of the old features, with wastes defined in lists and with requirements for plans, permits and inspections. But it introduced a new emphasis on prevention by requiring national waste reduction programmes by 2013. The waste hierarchy is to be followed in national legislation. Reuse is redefined, applying it to products and components that are not waste, and a new concept, 'preparing for reuse', covers processes by which waste products are prepared for reuse without reprocessing. This creates an additional stage in the waste hierarchy, splitting the concept of reusing products into two categories – those that have and have not become waste. There are new definitions of recycling, recovery and disposal. Incineration for municipal solid waste can be classed as recovery as long as the incinerators generate energy above a given efficiency threshold. The separate collection of paper, metal, plastic and glass must be in place by 2015, and by 2020 at least 50 per cent of these materials from households must be prepared for reuse or recycled. Extended producer responsibility requirements are introduced. Procedures are established for what constitutes a waste by-product, and when waste ceases to be waste. The collection and treatment of biowaste is promoted.

Table 6.1 and Figure 6.1 summarise the evolution of EU waste policy. Table 6.1 highlights the ideas put forward in the three waste strategies.

Figure 6.1 shows the legislation adopted during the four phases into which EU waste policy can conveniently be divided. The first phase (1975–85) covers the period before the 1989 strategy. All the six Directives adopted during that pioneering phase were replaced or amended during the second phase. The start of the second and third phases is marked by the second and third strategies. In the fourth phase several Directives were combined. Only the legislation abutting the right-hand edge of Figure 6.1 is currently in force.

TABLE 6.1 Waste strategies in the EU

Strategy	*Description*
1st	**September 1989 SEC(89)934 – Council Resolution May 1990 OJC 122 18.5.90**
	Waste to be considered under three headings:
	– prevention (clean technology, clean products)
	– recycling and reuse
	– optimisation of final disposal
	Disposal at 'nearest possible centres' preferred thus reducing movement
	'Priority waste streams programme' initiated
2nd	**July 1996 COM(96)399 – Council Resolution February 1997 OJC 76 11.3.97**
	Marks a shift to resource recovery
	The 'waste hierarchy' – the headings of the first strategy in descending order of preference:
	– prevention – the phrase 'producer responsibility' given currency
	– recovery (of materials and energy, so broader than recycling)
	– safe disposal
	The principles of 'proximity' and 'self-sufficiency of waste disposal facilities' are named
3rd	**December 2005 COM(2005)666 – Thematic strategy on waste prevention and recycling**
	A life-cycle approach to policy-making
	New focus on waste prevention
	New collection and recycling targets
	Shift to materials-based approach instead of focusing on end products

	Introduced during Phase I (1975–85)	Introduced during Phase II (1986–96)	Introduced during Phase III (1997–2004)	Introduced during Phase IV (2005–present)
Waste framework Directive	75/442	91/156		2008/98
Waste oils	75/439	87/101		2008/98
Hazardous waste	78/319	91/689		2008/98
Polychlorinated biphenyls (PCBs)	76/403	96/59		
Transfrontier shipment of waste	84/631	R259/93		R1013/2006
Packaging	85/339 (beverage containers)	94/62	2004/12	
Incinerators		89/369 and 89/429 (municipal); 94/67 (hazardous)	2000/76	2010/75
Batteries and accumulators		91/157		2006/66
Landfill			1999/31	
End-of-life vehicles			2000/53	
Waste electrical and electronic equipment (WEEE)			2002/96	2012/19
Restriction of hazardous substances in WEEE			2002/95	2011/65
Mining waste				2006/21

FIGURE 6.1 Waste legislation adopted 1975–2015
R = Regulation; all others are Directives.

Strategy on natural resources

Although an increase in recycling had been an aspiration from the beginnings of EU waste policy, a first priority had been the setting of standards for disposal. The main driver for the initial shift to an emphasis on resource recovery, set out in the 1996 waste strategy, was the need to reduce waste for disposal as outlets abroad were cut off and as existing disposal facilities came under greater pressure. Success in reducing emissions to air and water from industrial processes was leading to the production of more solid waste, and one justification for the 'integrated pollution and prevention control' Directive 96/61 (see Chapter 8) was the need to encourage 'clean production' that minimised total emissions, including solid waste, from industrial plants. At the same time, 'producer responsibility' led to products being better designed to facilitate recycling, partly stimulated by the 'priority waste streams' project and the resulting legislation. The Commission nevertheless felt the need to place resource recovery on a higher plane to contribute to sustainability, and accordingly the Sixth Environmental Action Programme of 2002 called for a thematic strategy on resources, in addition to the one on waste prevention and recycling discussed above.

After three years of debate and consultation, the 'Thematic Strategy on Sustainable Use of Natural Resources' (COM(2005)760) was published in 2005. It was initially envisaged that the Strategy would set quantitative targets 'for resource efficiency and diminished use of resources', but these proved impossible to set given the state of knowledge. An 'integrated product policy' (COM(2003)302) had already been initiated in 2003 as part of the EU's sustainable development strategy, and a number of other initiatives followed. These include the 'Raw Materials Initiative' (COM(2008)699) of 2008, the 'Action Plan on Sustainable Consumption and Production' (COM(2008)397) also of 2008, and the 'Roadmap to a Resource Efficient Europe' (COM(2011)571) of 2011 (see also Chapter 15).

The Seventh Environmental Action Programme of 2013 took this further. It was entitled 'Living well, within the limits of our planet', and in setting a long-term vision it spoke of a 'circular economy', a subject that had been attracting increasing attention:

> In 2050, we live well, within the planet's ecological limits. Our prosperity and healthy environment stem from an innovative, circular economy where nothing is wasted and where natural resources are managed sustainably, and biodiversity is protected, valued and restored in ways that enhance our society's resilience. Our low-carbon growth has long been decoupled from resource use, setting the pace for a safe and sustainable global society.

The origins of the concept of a circular economy have been described elsewhere (Hill 2015). In order to move towards such a circular economy, the Seventh Programme promised a review of existing product and waste legislation. It also called for the adoption of targets in a number of areas including resource efficiency. This led the Commission in 2014 to issue what became known as the 'circular economy package' including a Communication called *Towards a Circular Economy: A Zero Waste Programme for Europe* (COM(2014)398) and proposals to amend several existing waste Directives.

One of the early acts of the Commission that took office in November 2014, under its President Jean-Claude Juncker, was to withdraw the circular economy package as being inimical to business. Not long afterwards there was a backlash under the influence of progressive business leaders which led to the Commission announcing that a more ambitious package was to be reintroduced. At the time of writing this has yet to happen. Crucial to the successful reintroduction of the package will be the need to show that it will promote economic development and create jobs, as well as having environmental benefits.

Notes

1 For regulatory purposes a definition of waste is needed which provides certainty. To define waste as materials that have no value is open to the objection that there are times when a particular commodity, recycled paper for example, commands a price, and times

when no-one will buy it because there is a surplus. The answer to the question whether used paper is 'waste' would thus vary over time.

2 The European Free Trade Association is now composed of four non-EU states: Iceland, Liechtenstein, Norway and Switzerland.

3 No-one foresaw the fall of the Berlin wall so the Walloon decree cannot have been a response to the prospect of waste lorries arriving from Germany that might previously have gone to the GDR. The French decree, on the other hand, may well have been a response to the prospect of German waste, or indeed Flemish waste that could not go to Wallonia. In the latter case it could be seen as a kind of retaliation.

4 In Chapter 1 the argument was advanced that EU environmental policy began moving from the margins to centre stage in 1987. The closure of the Schoenberg waste disposal site in 1990 can be said to mark the shift of EU waste policy from waste management to the longer-term issue of resource recovery.

5 I once attended a conference where the participants were discussing EU legislation on climate change but did not know of the 'landfill' Directive and the contribution it made to reducing emissions of methane. Climate change, apparently, was one subject, waste another. This illustrates the difficulty of seeing EU environmental policy as a whole – see Chapter 1.

References

Hill, J (2015) The circular economy: from waste to resource stewardship, *Waste and Resource Management* 168(1), pp 4–14.

7

CHEMICALS – THE CINDERELLA OF ENVIRONMENTAL POLICY

One of the earliest items of EU environmental legislation established a scheme for testing new chemicals. Despite the originality and success of the scheme, chemicals policy was for a long time much less well known about than other environmental subjects, for several reasons: it was a new policy area; many effects of chemicals are invisible and only manifest themselves in the long term; the scientific basis on which controls are exercised are difficult for the layman to understand; the legislation and the language in which the subject is discussed are both fairly impenetrable; and the avenues for action by concerned citizens are not straightforward.

The effects of chemicals can be either acute or chronic, with acute effects being fairly obvious, as anyone will know who has spilt caustic soda when using it for washing or unblocking drains. It is the long-term effects, particularly of chemicals that persist in the environment and that can bioaccumulate, that are the focus of most concern.

Chemicals policy began to be better known in the 1990s for two main reasons. While it had long been clear that certain chemicals cause cancer, the work of the legendary Theo Colborn resulted in the publication in the USA of the 'Wingspread Statement' in 1991 (Colborn et al. 1996). This brought together information about birth defects, sexual abnormalities and reproductive failure in wildlife, and showed that some synthetic chemicals were capable of disrupting the endocrine system, with possible profound consequences given the crucial role that hormones play in controlling reproduction and development of both animals and humans. As a result, organisations concerned with protecting wildlife became much more engaged. Secondly, Sweden joined the EU in 1995 and its policy of protecting the Baltic Sea, where persistent chemicals accumulate, coupled with the fact that Sweden has no large chemical industry, had long made it an active proponent of more effective regulation of chemicals.

The need to review existing EU chemicals legislation was recognised by 1998, and in 2001 the Commission published a White Paper setting out ideas for reform and consolidation. After controversial negotiations, which had the effect of making the subject much better known to politicians, the EU in 2006 adopted Regulation 1907/2006, called REACH (Registration, Evaluation, Authorisation and Restriction of Chemicals). During these negotiations an international workshop was held in Paris, at which I was invited to provide a brief history of existing EU legislation (Haigh 2005). As the length of the paper was strictly limited I could only give a condensed account. My theme was that REACH should be seen as evolutionary rather than revolutionary. Many commentators, including some from the Commission, were saying that existing chemicals legislation had failed when they could more accurately have said that some parts had failed. In a curious inversion of the normal desire to celebrate success, emphasis was instead placed on failure in order to justify what was said to be a fresh start.

A brief history of EU regulation of chemicals – paper delivered in Paris, June 2005

Character of chemicals policy

In all EU Member States chemicals policy is dominated by EU legislation more completely than any other branch of environmental policy. There are two reasons for this. Unlike air or water pollution, or nature protection, which have long established national traditions, EU chemicals legislation developed before any Member State had developed a strong tradition of its own. Chemicals policy is also largely about regulating the sale of individual chemicals, and the EU cannot allow national policies to diverge without fragmenting the common market whose creation and preservation has always been a main task of the EU.

Two other preliminary comments need to be made about chemicals policy. First, until recently the subject was little known outside the chemical industry and the limited circle of toxicologists. Only with the new found attention to endocrine disrupting substances did it catch the public's attention. Secondly, if we consider traditional air and water pollution policy to be about controlling emissions of chemicals in the form of waste, chemicals policy is about controlling the use of chemicals before they become waste. Chemicals policy therefore embodies the preventative and precautionary principles.

This paper deals only with industrial chemicals and not with pesticides and pharmaceuticals, which are subject to separate EU legislation. These are intended to be dangerous, for example to kill pests. Industrial chemicals, by contrast, are used because they are useful despite sometimes being dangerous.

The four phases of EU chemicals policy

Chemicals policy can be said to have started seriously in 1973 when the Organisation for Economic Co-operation and Development (OECD) Council issued a Decision requiring its member countries to regulate the use of polychlorinated biphenyls (PCBs). This was in response to a number of incidents including poisoned rice oil in Japan and spectacular bird deaths in the Irish Sea. Despite the binding character of an OECD Council Decision, the EU responded by proposing a Directive to make the OECD Decision uniformly effective in all Member States. It also went further by creating a framework for restricting the marketing and use of any dangerous substance. This was the beginning of four overlapping phases into which EU chemicals policy can be divided. The proposed comprehensive reform of EU legislation known as REACH is best understood as evolving out of the three earlier phases.

- **1970s** – Ad hoc restrictions on the marketing and use of chemicals that were known to be harmful (often following tragic accidents).
- **1980s** – A systematic and proactive approach to new chemicals which were not allowed onto the market before they had been tested.
- **1990s** – A programme for evaluating existing chemicals. Priority lists totalling 140 chemicals have so far been agreed. Of these only a few have so far been evaluated and fewer restricted.
- **2000s** – A proposed consolidation and extension of the earlier phases intended in particular to increase knowledge of the thousands of existing chemicals more quickly than at present, and to provide a system that will ban the use of all chemicals 'of very high concern' unless they are expressly authorised. The proposal is known as REACH.

There was a yet earlier phase in the 1960s, when the EU adopted a Directive on the classification, packaging and labelling of chemicals. This was before the EU had an environmental policy and it then merely harmonised the labelling requirements introduced by some Member States to protect workers. However, the regime for testing new chemicals (phase 2) was adopted as an amendment of this labelling Directive 67/548.

Terminology

The language in which chemicals policy is discussed makes important distinctions that are often confusing to the non-expert. I therefore offer the following explanations in the knowledge that toxicologists may find them oversimplified.

The word 'chemical' is often used loosely to cover both 'substances' and 'preparations' which are defined more precisely: 'substances' are chemical elements and their compounds, whereas 'preparations' are mixtures or solutions of two or more substances (e.g. paints, inks, solvents).

'Hazard' is a property intrinsic to a chemical substance, for example its toxicity, flammability, corrosivity or carcinogenicity. 'Risk', on the other hand, relates to the likelihood of harm and so depends on exposure, which in turn depends on the uses to which a chemical is put. 'Hazard assessment' accordingly means the identification of the adverse effects which a chemical has the capacity to cause. It is a scientific process which may involve tests on laboratory animals. 'Risk assessment', on the other hand, is a more difficult process which starts from hazard assessment but also involves exposure assessment. Since information on use and exposure is often limited (and non-existent in the case of new chemicals), risk assessment is subjective and requires expert professional judgement. This distinction is important. For example, the OECD programme on existing chemicals results in the provision of information in the form of hazard assessments, whereas the EU existing substances Regulation goes further in requiring more time-consuming risk assessments before restrictions can be imposed.[1]

'Risk management' or 'risk reduction measures (or strategies)' are phrases used to describe any one of a number of practical steps following hazard or risk assessment. These can range from a total ban to a mere warning label, and can include the provision of detailed advice on 'safety data sheets'; restrictions on marketing and use; controls over emissions; setting environmental quality standards; and instituting surveillance programmes.

A 'downstream user' is an industrial user of a chemical, other than the manufacturer or importer, for example a paint maker. A consumer is not a 'downstream user'. There are many thousand times more downstream users than manufacturers. Until now, downstream users have not been responsible for contributing to hazard and risk assessments, but they will be more involved under REACH, which is one reason why it is controversial.

Phase 1 – Restrictions – starting 1978

Directive 76/769 enables restrictions to be placed on the marketing and use of any dangerous substance or preparation. It cannot ban the production of a substance, and when the Montreal Protocol on the ozone layer required a production ban on chlorofluorocarbons (CFCs) a separate EU Regulation was adopted.

The Directive initially restricted only three chemicals. PCBs could only continue to be used in closed-system electrical equipment and it was not till nine years later that most uses of PCBs were banned. Over the years the Directive has been amended to impose restrictions on many other substances, including asbestos; lead in paint; marine anti-fouling paint; cadmium; fire retardants; carcinogens; creosotes; some cements; and chlorinated solvents. Some of these restrictions have been controversial and involved disputes with countries outside the EU, the best example being asbestos.

A few of the restrictions have followed risk assessments carried out under the regimes for both new and existing chemicals (see Phases 2 and 3 below).

Under REACH the provisions of Directive 76/769 will be continued in a modified form: it will also be possible to ban the manufacture of a chemical. In addition REACH will ban all chemicals 'of very high concern' unless they are authorised.

Phase 2 – New chemicals – starting 1981

In the early 1970s a debate developed about the need for an 'early warning' or, as we would now say, a precautionary system for new chemicals, and this found expression in the EU's first action programme on the environment of 1973. This called for controls over new chemicals before being marketed. In the USA this debate led to the Toxic Substances Control Act 1976 (which also dealt with existing chemicals) and in the EU to Directive 79/831 (which amended for the sixth time the classification Directive 67/548).

Directive 79/831 was highly original and, with various subsequent amendments, laid down many of the principles of EU chemicals policy. Some of these had been developed within the OECD. It made a distinction between 'new' and 'existing' chemicals. All chemicals are 'new' unless they are listed in the European Inventory of Existing Commercial Chemical Substances – EINECS – as having been on the EU market before September 1981. Over 100,000 are listed.

Since 1981 a manufacturer of a 'new' chemical substance has had to submit the results of tests sufficient to evaluate possible harmful effects, and its assessment of the results, to the competent national authority. With increasing production more information has to be submitted. The authority sends the information to the European Commission, which sends it to the authorities in all other Member States. Any of these can make enquiries. If no objections are raised within 60 days the manufacturer has assured access to the whole EU market.

Nearly 3,000 chemicals have been notified since the scheme began, and anecdotal evidence suggests that some chemicals which manufacturers began developing have never been marketed because they were found to be more dangerous than expected. This is the precautionary principle at work, although the authorities will not have been told. The testing will also have produced information which can enable the chemicals to be used more safely. For these reasons the scheme is thought to have worked well.

The chemical industry supported the Directive when it was proposed – in contrast with REACH today. They could see that a single European system was preferable to different testing regimes that might develop in different Member States. Negotiations on the Directive also coincided with the drafting in the USA of rules to implement the Toxic Substances Control Act 1976. The European Chemical industry was fearful that these rules might discriminate against European exports to the USA and wanted a good EU regime so that the EU could negotiate from strength with the USA if necessary. They could see that the size of the EU market to US manufacturers would mean that the European Commission could negotiate more effectively than Germany, France or the UK negotiating separately. The Directive is thus an example of synergy between environmental and trade requirements.

Phase 3 – Existing chemicals – starting 1993

The regime for evaluating existing chemicals developed slowly, not surprisingly given the difficulty of the subject, industrial resistance and the reservations of some

Member States. The need for more information on the thousands of existing chemicals had been recognised in the 2nd, 3rd and 4th Environmental Action Programmes (of 1977, 1983 and 1987), but it was not till 1990 that the Commission felt able to make a proposal. This was adopted in 1993 as Regulation 793/93, known as the Existing Substances Regulation (ESR). It is because of disappointing progress with ESR that the pressure has grown for REACH.

Briefly, ESR requires manufacturers to send only existing data to the Commission. The Commission then draws up priority lists of chemicals needing attention, and work on risk assessment is shared between the Member States. The authorities can then propose risk reduction strategies. More fully these steps are as follows.

Data reporting: Manufacturers submit data – but only what already exists – relevant for an evaluation of risk to the European Chemicals Bureau (ECB). The ECB – established within the EU's Joint Research Centre at Ispra, Italy – manages the International Uniform Chemical Information Database (IUCLID). More information has to be submitted for 'high-volume' production chemicals (more than 1,000 tonnes per year).

Priority setting: Using the submitted data, the Commission draws up lists of priority chemicals requiring immediate attention taking into account certain criteria. By 2000 four lists had been adopted totalling 140 chemicals.

Risk assessment: Each priority chemical is allocated to a Member State which designates a 'rapporteur' to evaluate the chemical. If a 'base set' of data is not available, the manufacturer must carry out the necessary testing and submit this to the rapporteur within 12 months. The rapporteur evaluates the information and decides whether the manufacturer is to be required to supply further information or carry out further testing. A Committee can decide whether supplying this further information is to be obligatory, and the time limits, but this can only be done when there are valid reasons for believing that a chemical presents serious risks. The rapporteur then carries out the risk assessment and sends conclusions to the Commission.

Risk reduction: The rapporteur's conclusions can suggest a strategy for limiting risk. Any proposed restriction on marketing and use must be accompanied by an analysis of the advantages and drawbacks of the chemical and the availability of replacement chemicals. The recommended strategy can be adopted by the Committee and published. If there are to be restrictions, these can be proposed by the Commission under Directive 76/769 (see Phase 1).

By early 2005 only seventeen risk reduction strategies had been published and only a few chemicals had been restricted. No-one publicly foresaw such slow progress. Industry criticises governments for not providing adequate resources for carrying out risk assessments. Others say that industry is slow to provide information and can use many delaying tactics to gain an extended market period. Without reasons for believing that a chemical poses a serious risk the authorities cannot demand information, and without the necessary information it is difficult to provide the reasons.[2] Unlike new chemicals, where it is in the interest of the manufacturer to supply information – since without it marketing is not allowed – with existing chemicals it is not in the interest of the manufacturer to supply information.

Phase 4 – REACH – starting 2009?

REACH is intended to overcome the criticisms of the ESR by placing much more responsibility on manufacturers and downstream users to provide useful information about the thousands of chemicals on the market. It also abolishes the distinction between 'new' and 'existing' chemicals; introduces an authorisation system for chemicals 'of very high concern', for example those that are very persistent and very bioaccumulative; and replaces the ECB with a much larger European Chemicals Agency. It is the longest and most complicated item of environmental legislation to have been proposed so far by the European Commission. It is also the most controversial, the question being whether sufficient useful information can be provided without excessive burdens being placed on the European chemical industry and without excessively increasing the amount of animal testing.

Paper delivered June 2005

Developments since 2005

The Commission published a White Paper in 2001 on a 'Strategy for a future chemicals policy', and then, before formally proposing REACH, it also took the unusual step of publishing a draft for consultation in May 2003. This resulted in an unprecedented reaction from three Member States – France, Germany and Britain. A letter to the Commission President, signed by President Chirac, Chancellor Schröder and Prime Minister Blair, warned of the possible impacts REACH might have on the competitiveness of the European chemicals industry. Since these three Member States could at that time command a blocking minority of votes in the Council, the warning could not be brushed aside and influenced the decision that the proposal should be examined by the Competitiveness Council (composed of industry Ministers) as well as the Environment Council. From then on all parties to the negotiations were on notice that these three Member States were likely to coordinate their positions in order to influence the outcome. In the European Parliament both Committees also examined REACH, with dialogue between them, but with the Environment Committee in the lead. The battle lines were thus drawn for the protracted negotiations: industrial versus environmental interests. This was partly reflected in dialogue between the Council and Parliament, with the competitiveness of the European chemical industry being uppermost in the Council's mind and the Parliament being more sympathetic to environmental interests. Unlike Directive 79/831 on testing new chemicals, which had been supported by the European chemicals industry, REACH was fiercely opposed. The extra burdens placed on industry to supply information was one reason. Another was that many more industrialists would be affected since 'downstream users' would also be required to supply information for the registration process. Compared with the published draft, the formal proposal issued in October 2003 reduced the burden on industry, and it was to be modified again before being finally adopted two years later during the British Presidency. The Presidency always has the task of looking for a compromise to secure agreement and can therefore play an influential role.

Two issues dominated the negotiations: the amount of information that had to be supplied by a manufacturer/importer when registering a chemical, and the authorisation procedure for chemicals of very high concern.

All chemicals have to be registered with the European Chemicals Agency (ECHA), a newly created body located in Helsinki, accompanied by a safety report based on a safety assessment. During the negotiations, industry wanted to minimise the amount of information that had to be supplied while environmental interests insisted that enough information had to be supplied to enable potentially harmful effects to be identified. Just before the first reading in the Parliament in November 2005, a compromise was agreed between the two main political groupings, the centre-right European People's Party (EPP) and the socialists. The Parliament's rapporteur explained that, though the compromise may not have provided the best balance, it was 'the best politically available balance'. It remains an open question whether some of the many thousands of chemicals now registered will in time prove to be more dangerous than anticipated.

Debate continued for longer over the authorisation procedure required for chemicals of very high concern. In December 2005 the Competitiveness Council agreed a 'common position' that did not accept the Parliament's view that substances of very high concern could be authorised only when there are no suitable alternatives. It would allow such substances to be authorised in certain circumstances if there was an 'adequate control' of risks. The Council, however, did not allow authorisation on the grounds of 'adequate control' for substances that are persistent, bioaccumulative and toxic (PBT) or very persistent and very bioaccumulative (vPvB). At its second reading the Parliament maintained its earlier position, but after intense negotiations between Parliament, Council and Commission, the Parliament had to move from a requirement for mandatory substitution to mandatory substitution plans.

The authorisation process starts with a chemical of very high concern being put on a list known as the candidate list of substances for possible inclusion in the Authorisation List. The authorisation procedure itself entails setting a date when a chemical will be banned unless authorised. This gives industry a clear idea of which chemicals are problematic and time to find safer alternatives. At a conference organised by ECHA in February 2015, it was revealed that no application for authorisation had been made by the due date for half of the chemicals on the Authorisation List so that these chemicals can no longer be marketed. Furthermore, some of the applications were for a short period only, to give more time for substitution to occur. ECHA, while noting scope for improvement, concluded that the authorisation system is working and putting pressure on industry to substitute chemicals with safer alternatives. However, there is some disquiet about how this authorisation procedure is working in practice. Environmentalists consider that certain authorisations for use should not have been granted, and some in industry complain about the costs and uncertainties. The 'precautionary principle' has been important for chemicals policy, and some examples of chemicals restricted or banned, where the principle has been invoked, are given in Chapter 13.

One area that remains problematic is the regulation of endocrine-disrupting chemicals (EDCs). When REACH was being negotiated, agreement could not be reached as to whether such substances should be authorised for use if there was adequate control of risks, or only if the socio-economic benefits of use outweigh the risks and there were no suitable alternatives. A review clause was accordingly put into REACH with a deadline of June 2013 for determining this. Due to disagreements among scientists as to whether EDCs should be considered to have thresholds for effects, coupled with an industry backlash against the costs of controlling EDCs, this deadline was missed. So, too, were other deadlines for criteria to identify EDCs in legislation relating to pesticides and biocides. It was not until mid-2014 that the Commission presented its finalised review paper on the conditions that must be met in order for an EDC to be authorised under REACH. This paper suggested that authorisation of those substances put on the candidate list for their EDC properties could only be authorised via the adequate control route if it could be demonstrated that a threshold existed. Far from being a clear-cut decision, in each case arguments are likely to ensue as to whether safe levels can actually be defined.

During the registration process, the manufacturer/importer must prepare a safety assessment for each chemical which, among other matters, must say whether the chemical is PBT or vPvB. The 'safety assessment' has similarities with the 'risk assessment' required by the previous legislation (see above) but is less rigorous, perhaps due in part to an inevitable reluctance on the part of industry to indict their own chemicals. Downstream users must supply information on how they use the chemical so the manufacturer can compile the assessment. Previously many manufacturers did not always know how their chemicals were being used, and one of the successes of REACH has been to create a flow of information up and down the supply chain about the properties of chemicals so that they can be used more safely. A major problem nevertheless is the quality of the safety assessments. ECHA can evaluate the registration dossiers and evaluate substances. At least 5 per cent of dossiers must be evaluated for completeness, and many have been found deficient. ECHA also compiles three-year rolling action plans for evaluating substances on the grounds that the dossiers, or any other available information, suggest that they are a risk to human health or the environment. Substance evaluation goes beyond checking whether the dossier is complete and is an evaluation of whether the use of that substance poses a risk.

Compared with the legislation it replaced, REACH places more responsibility on industry to assess the risks of existing chemicals. The effectiveness of the previous regime for testing new chemicals depended on a manufacturer not being able to market the chemical until it had been assessed. The failure of the previous regime for existing chemicals was the result of the manufacturer being able to continue to market the chemical while it was being assessed and having every interest in delaying the assessment. REACH promises to apply the successful feature of the old regime for new chemicals to all chemicals. This is embodied in Article 5 headed 'No data, no market' (perhaps the first time a slogan devised by an NGO – Greenpeace – has entered EU legislation)[3]. Article 5 denies a market to any chemical not properly registered. What should be a serious sanction has yet to be applied despite the fact that many registrations have been shown to be seriously deficient. The slogan will remain just that until a chemical is removed from the register for serious deficiencies in its registration dossier.

Another unprecedented aspect of the adoption of REACH was the joint statement of opposition that was issued in June 2006 by a number of countries, led by the USA. The signatories were among the EU's biggest trading partners, including Australia, Brazil, India, Japan, Korea, Malaysia, Mexico, Singapore, South Africa, Thailand and USA. Whether the statement added anything to the European debate about how best to have a flourishing and forward-looking chemicals industry that does not cause long-term damage to humans and wildlife must be doubted. Certainly it did not deter the EU from adopting the Regulation, though it might have deterred individual countries had they been acting on their own. What it did show was that the world was watching. Now that REACH is in place, a manufacturer anywhere in the world exporting to the very large EU market must conform to REACH, which has therefore become the benchmark for chemicals regulation worldwide.

Notes

1 I have attended meetings at which participants discussed 'assessments' of chemicals. It became apparent that different people meant different things by the same word.
2 At the workshop this was described as the 'Catch 22' that resulted in the failure of the ESR.
3 The phrase 'no data, no marketing' was used in a checklist of demands for reform drafted by Axel Singhofen of Greenpeace presented at a European Commission brainstorming meeting on chemicals held on 24/25 February 1999. This was simplified in later Greenpeace papers to the catchier phrase 'no data, no market'. The concept of course was already known to many. By June 2001 the catchier phrase was being used in a report prepared by the Netherlands Government.

References

Colborn, T, Myers, J and Dumanoski, D (1996) *Our Stolen Future – How Man-Made Chemicals are Threatening our Fertility, Intelligence and Survival*, London: Little, Brown.

Haigh, N (2005) A brief history of EU regulation of chemicals. In: Weill, C, ed., *European Proposal for Chemicals Regulation: REACH and Beyond*, IDDRI Studies No. 03/2005, Paris: Insistut du développement durable et des relations internationales.

8

INTEGRATING POLLUTION CONTROL

Pollution first became an important subject for public policy as towns grew in size, as industries proliferated, and as the causes of diseases such as cholera were recognised. The problems of rivers, of air and of solid waste were initially perceived as different subjects, and different institutional arrangements evolved for the different environmental media (air, water and waste on land). The constitutional, cultural and geographical context shaped this evolution, and different ways of thinking took root and became embedded in legislation. As we have seen in Chapters 4–6, the EU also began by adopting separate legislation for the different environmental media. These were the responsibility of separate units within what is now the Commission's Directorate-General for the Environment.

In the early 1970s, Ministries for the environment were being created in several countries and this stimulated thinking about how to bring the separate subjects together. The need to control the use of potentially dangerous chemicals had also emerged in the 1970s as an important new subject (Chapter 7) resulting in the creation of yet new administrative units. By the mid-1980s discussions on what was variously called 'cross-media' or 'multi-media' or 'integrated' pollution control were taking place internationally within the Organisation for Economic Co-operation and Development (OECD). The Chemicals Group of the OECD Environment Directorate took a particular interest as it was working on comprehensive risk management of chemicals and it was evident that synthetic chemicals could reach vulnerable receptors via all the environmental media. As a result of a chance meeting at an OECD workshop on 'integration' in 1985, the Institute for European Environmental Policy (IEEP) and the Washington-based Conservation Foundation (CF)[1] began a joint project conducted between 1986 and 1988. The resulting book, *Integrated Pollution Control in Europe and North America* (Haigh and Irwin 1990) included case studies of moves towards integration in several countries. To simplify a difficult subject, we relied on an insight that CF had already formulated. CF had seen 'integrated pollution control' as a broad concept best understood as involving a shift in focus away from the three media: 'water', 'air' and 'land', and instead to 'substance', 'source' and 'region', each of which offered scope for practical action.[2] We never doubted, however, that focusing on an individual medium was the best approach in many circumstances. The project had some influence, even beyond the OECD and EU, and in 1996 the EU adopted Directive 96/61 on 'integrated pollution prevention and control' (IPPC), although confined to industrial plants.

So it was that I was invited to present a paper to the 1997 annual conference in Berlin of the German Council of Environmental Lawyers at a time when implementing the IPPC Directive 96/71 was under discussion (Haigh 1998). The Directive was contentious in Germany, not just because it introduced new concepts, but because of Germany's federal structure which made the Directive particularly tricky to implement. Under the German Basic Law, air pollution policy is a matter reserved for the Federal Government whereas water policy is shared between the Länder and the Federal Government.

Given the formal nature of the proceedings, and that I was allowed to speak in English, and also as my message challenged a deeply held German preference for the certainty provided by centrally fixed numerical standards for emissions to air and water, my paper was deliberately dry in tone and relied heavily on quoting official texts. Its title was set by the conference organisers. A more personal account of the development of the Directive appears in the Appendix, which describes the CF/IEEP project.

Integrating environmental laws – existing laws, aims, possibilities – paper delivered in Berlin, October 1997

'Internal' and 'external' integration

In the Netherlands, where they have thought about these matters for many years, a distinction is made between 'internal integration' and 'external integration' of environmental policy and law.

By 'internal integration' they mean integrating policies for pollution of water and air and land. By 'external integration' they mean integrating environmental considerations into other policy areas such as agriculture, transport and energy. 'External integration' is therefore concerned with much more than pollution.

The two forms of integration are, of course, not entirely separate since agricultural activities result in both air and water pollution, and attempts to limit water and air pollution will not be successful without supportive agricultural policies. The distinction is nevertheless useful. I am told that the terminology arose from the fact that air and water pollution are both internal to the Environment Ministry, whereas external integration involves dealing with other Ministries (agriculture, transport etc.), which is altogether a more difficult matter, as anyone who has had experience of bureaucracies knows well.

This paper is confined to 'internal integration' or integrated pollution control, which is itself a difficult enough subject. It will concentrate, though not exclusively, on developments in the UK and the EC. EC legislation, particularly the IPPC Directive 96/61, is now causing changes in Germany, and while it is not correct that the UK inspired the Directive as many people in the UK believe, there is no doubt that UK legislation had an important influence on the shape of the Directive. But let me start in the USA.

Permit me first one further remark on terminology. The phrase integrated pollution control (IPC) is sometimes extended to integrated pollution prevention and control (IPPC) to emphasise that the objective is to *prevent* pollution. The two phrases mean the same if control of pollution is interpreted, as it must be, to mean its prevention as well as its minimisation.

Disappointment in the USA

The creation of the Environmental Protection Agency (EPA) in the USA in 1970 was an attempt at 'internal integration'. The reasons were well expressed by President Nixon – not generally remembered as a great environmentalist – when he proposed to Congress the creation of the EPA:

> Despite its complexity, for pollution control purposes the environment must be perceived as a single inter-related system. Present assignments of departmental responsibilities do not reflect this inter-relatedness. Many agency missions, for example are designed primarily along media lines – air, water and land. Yet the sources of air, water and land pollution are inter-related and often interchangeable ... a far more effective approach to pollution control would identify pollutants; trace them through the entire ecological chain; ... determine the total exposure of man and his environment; examine interactions among forms of pollution; and identify where in the ecological chain interdiction would be most appropriate.
> *(quoted in Davies 1990)*

Despite the declared intention to practise integration, the US EPA nevertheless very soon found itself working along single medium lines because Congress insisted on passing separate laws on water and air with different targets to be met by different deadlines. The EPA responded by setting up different air and water departments to meet those deadlines. The US experience teaches us that a single administrative agency will not necessarily operate in an integrated way unless you have legislation requiring it to do so. Such 'integrated' legislation is still missing in the USA. Despite a number of attempts to achieve integration, the original promise of the EPA has not been fulfilled.

The OECD Council Act of 1991

The OECD had absorbed the lesson from the USA when in 1991 it adopted a Council Recommendation on 'integrated pollution prevention and control'

(OECD 1991). This remains the fullest and clearest official statement describing integrated pollution control. It calls on the Member Countries of OECD:

> to practise integrated pollution prevention and control, taking into account the effects of activities and substances on the environment as a whole, and the whole commercial and environmental lifecycles of substances when assessing the risks they pose and when developing and implementing controls to limit their release.

That broad statement is followed by more precise recommendations. Member Countries are to ensure that their laws and regulations support integration by:

> evaluating the extent to which they present impediments to the implementation of an integrated approach.

They are then (a) to amend their existing laws and regulations to remove such impediments, and (b) to adopt new laws and regulations to promote integrated pollution control.

All EC Member States are now having to do this in order to implement the IPPC Directive 96/61. The point about getting the right legislation having been recognised, the recommendation goes on to say that Member Countries shall:

> adopt administrative procedures and institutional measures to ensure that an integrated approach to pollution prevent and control can be achieved efficiently.

In other words, you need both the **right institutions** and the **right legislation**. The USA, as we have seen, may have the first but not the second.

The preamble to the OECD Act provides as neat a justification for integrated pollution control as I have seen in any official document. Three reasons are given in the considerata:

- considering that substances can move among environmental media (air, water, soil and biota) as they travel along a pathway from a source to a receptor and can accumulate in the environment;
- considering that controls over releases of a substance to an environmental medium can result in shifting the substance to another environmental medium;
- considering that in many Member countries, pollution control efforts focus on each environmental medium separately and that controls over the marketing and use of substances are carried out as separate activities.

The first two considerata concisely state the double recognition that is driving integration: that (a) the environment operates as an interconnected whole and polluting substances are not neatly contained within one medium but move between them, and (b) controls within one medium can simply result in redistribution to another medium.

The third of the considerata recognises that in most countries different people deal with emissions to water and to air, and that yet another set of people deal with approving chemicals and controlling traded products. Often they do not talk to each other.

The preamble then makes a statement of faith when it says that 'such separate efforts undertaken alone are not necessarily the most efficient and effective way to protect the environment'.

Integrated pollution control in the UK

Before 1987, air pollution, water pollution and waste disposal were regulated in the UK under separate laws by quite different authorities. Air pollution of major plants was the responsibility of a national inspectorate (originally created in 1863). Air pollution of smaller plants was the responsibility of lower-level local government – District Councils. Water pollution was the responsibility of river basin-based water authorities (created in 1974). Waste disposal regulation was the responsibility of higher-level local government – County Councils.

In 1976 the Royal Commission on Environmental Pollution published a report titled *Air Pollution Control: An Integrated Approach* (RCEP 1976). This noted that, because of the connections between different forms of industrial pollution, it made little sense to look at one aspect (e.g. air or water or waste) in isolation. It cited several examples of 'cross-media' problems where control in one medium created an extra burden in another. It therefore proposed a new, unified pollution inspectorate for major plants but did not propose a change in the law.

The Government first rejected the idea in 1982 but then changed its mind and, without changing the law, created a new unified pollution inspectorate in 1987 called Her Majesty's Inspectorate of Pollution (HMIP).[3] In 1988 it combined all the separate river basin-based water authorities into a single National Rivers Authority (NRA). In 1990 the Environmental Protection Act entirely modernised pollution control for major industrial processes under a system called 'integrated pollution control'. HMIP now granted a single authorisation for such processes covering discharges to air, water and the generation of waste. Where releases of prescribed substances to more than one environmental medium may be involved, 'best available techniques not entailing excessive costs' (BATNEEC) must be used to minimise pollution to the environment as a whole. Before granting the authorisation, HMIP had to consult the NRA to see if the proposed discharge would meet water quality objectives. If it did not, then HMIP had to reconsider the authorisation.

The phrase BATNEEC, which replaced the nineteenth-century phrase 'best practicable means', was derived from EC Directive 84/360 on air pollution, but is different in a crucial way. Under Directive 84/360, BATNEEC is to be applied *to prevent pollution of air.* Under the UK Act of 1990, BATNEEC is to be applied to *prevent pollution of the environment as a whole.* This should stimulate both the industrialist and the authority to think about the whole process rather than just adding end-of-pipe solutions, and should result in 'clean technology'.

To illustrate this important difference let us imagine a plant discharging both to air and to a small trout stream. To comply with EC Directive 84/360 the best technology to minimise air pollution must be applied. This could be a wet scrubbing system which results in discharges that would kill the trout. Let us call this technology A. Under the UK system a technology B would have to be chosen which resulted in a discharge that was acceptable in the trout stream, and also acceptable for air pollution purposes, even if it resulted in some more emissions to the air than technology A. The choice of technology will therefore depend on the site, and will require judgement on the part of the authority, which has to balance the need both to protect water and to protect air. The system will only work if an experienced and reliable inspectorate exists. It follows that integrated pollution control only works if you have the right *institutions* as well as the right *legislation.* You will appreciate that UK legislation could result in breach of EC Directive 84/360, which is one reason why the UK was so supportive of IPPC Directive 96/61 which repeals 84/360.

A further major development took place in the UK in 1995 with the creation of the Environment Agency. (It covers England and Wales: a separate Agency covers Scotland.) This has combined HMIP with the NRA and with the waste regulatory functions of local authorities. What used to be discussions between different authorities are now easier so that integration should be more effective.

Integrated pollution control in the EC

The idea of an 'integrated permitting Directive' was first proposed at a conference organised by my Institute in Brussels in November 1988 (Haigh and Irwin 1990). In 1989 we submitted a report to DG XI (environment) recommending such a Directive (IEEP 1989). Our report showed that some form of integrated permitting of industrial plants was being developed in several Member States and there was a risk that the EC would be left behind with old-fashioned single medium legislation. DG XI accepted our conclusions but drafting was delayed while the Eco-audit Regulation 1836/93 was developed. It only started in 1991 when a British official was seconded to DG XI. While British legislation was certainly one influence, it was not the only one.

Directive 96/61 takes its name from the OECD Council Act (OECD 1991). It differs from the UK legislation in several important respects (Emmott and Haigh 1996) but shares with it the crucial objective that the permit *is to achieve a high level of protection for the environment as a whole* (Article 9). For the reasons given above this may be inconsistent with the requirements of Directive 84/360 which will therefore be repealed.

Directive 96/61 uses the phrase BAT ('best available techniques') instead of BATNEEC used in 84/360, but as BAT is defined as taking into consideration costs and advantages [Article 2(11)] the difference is more symbolic than real.

BAT is also qualified in Article 9(4) so that emission limit values, while based on BAT, shall take account of the geographical location of the plant and local environmental conditions. BAT can therefore vary from place to place, and the problem raised by my trout stream example can be accommodated. I appreciate that Germany opposed this when the Directive was being negotiated and wanted uniform emission limits without the possibility of geographical variation. I believe that showed a misunderstanding of the concept of integrated pollution control since minimising harm overall must imply differing solutions in different locations.

One difference between Directive 96/61 and the UK legislation is that the UK legislation requires a single permit granted by a single authority. The Directive, however, allows more than one permit from more than one competent authority so long as they are coordinated in order to guarantee an effective integrated approach (Article 7). This was introduced specifically to accommodate, for example, Germany and the Netherlands, where there is more than one competent authority.

Some people, particularly in Germany, fear that the flexibility to allow different emission limits depending on geography (essential to the British system) can be abused, and some people in Britain believe that coordinating different competent authorities (essential for the present German system) does not guarantee proper coordination and is inefficient. Time will tell whether these fears are justified, but in order to know, there needs to be effective evaluation of the implementation of the Directive in all Member States. Since the Directive is likely to be implemented differently in different Member States, it is essential that we all exchange information and officials and so learn from each other. Germany needs to consider whether it should adapt its institutions as well as its legislation to implement the Directive.

With the adoption of the IPPC Directive, the EU is now well in advance of the USA, and people from outside the EU are showing great interest in the Directive.

Different approaches to integrated pollution control

The EC Directive 96/61 relates only to industrial plants and is only one embodiment of the much broader idea of integration set out in the OECD Council Act. The guidance that accompanied the OECD Council Act explains that an integrated approach effectively requires a shift in focus in ways of thinking. Instead of concentrating on the three media one at a time, it is useful to concentrate on three approaches: substances, source and region (OECD 1991).

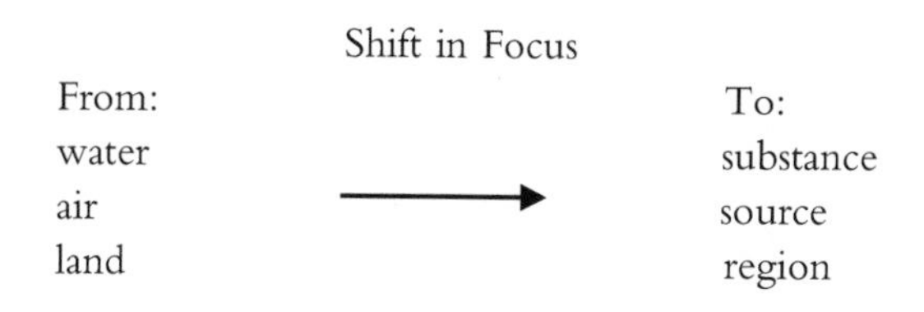

The regional (or geographical) approach

An example concerns the North Sea. Initial attempts to protect it concentrated on dumping from ships – the Oslo Convention of 1972. This was followed by the Paris Convention of 1974, which was concerned initially with discharges to water from land-based sources including rivers. When it was realised that a significant proportion of inputs of some substances to the North Sea came from aerial deposition, the Convention was amended. It is now being replaced by a new OSPAR Convention [Convention for the Protection of the Marine Environment of the North-East Atlantic] 'which addresses all sources of pollution and adverse effects of human activities'.

Another example is provided by the Great Lakes in North America, where pollution arises not only from direct discharges to water and from atmospheric depositions, but also from leachate from the land.

Effective environmental protection of these regional water bodies requires attention to all routes by which pollutants travel. The concept can of course be applied to small geographical areas as well as to large ones.

In the UK, the Environment Agency is adopting the regional approach by preparing Local Environment Agency Plans (LEAPs). These are an extension of the previously existing river catchment plans which concerned all aspects of water (quality and quantity) and which are now being widened to consider air and land as well.

The substance approach

In theory it should be possible to trace the path of a chemical substance through its whole commercial and environmental life cycles and assess where it poses risks. Appropriate controls can then be applied. An example is CFCs, which are relatively non-toxic in use but eventually destroy stratospheric ozone. The response has been to ban production.

Another example is the EC Directive 92/32, which requires all new chemical substances to be evaluated before they are marketed. The evaluation is to take account of risks in all media, taking account of likely uses, and once these risks have been evaluated the marketing and use of substances can be restricted as necessary.

Another example of the substance approach is the EC eco-label scheme for products with reduced environmental impact (Regulation 880/92). The impact of products is to be assessed on all environmental media throughout the entire life cycle of the product (pre-production, production, distribution, use and disposal).

The source approach

This is the approach of the IPPC Directive 96/61 discussed above.

Possible future developments

Integrated pollution control is a broad concept that should not be regarded as confined to industrial plants as in the EC Directive. It is a dynamic concept capable of considerable extension.

One possible development of the source approach related to industrial plants is the use of mass balances, that is, considering all the inputs into an industrial process (raw materials, energy, water etc.) and all the outputs, whether in the form of emissions from pipes or stacks, or fugitive emissions, as well as in the form of products. As a legal requirement that is still something for the future, although some experiments have been conducted. However, one can see a move in that direction. The UK law allows the authorisation to include control on inputs of substances as a means to control outputs of pollutants. The EC Directive goes a little further still, as it aims to reduce emissions regardless of harm in pursuit of waste reduction and conservation of resources. This applies particularly to energy and water. An emissions release inventory is foreseen.

The other approaches to integrated pollution control, the substance and regional approaches, are also capable of development. The recent proposals from the Swedish Chemicals Committee suggest that we can expect a greater integration between 'chemicals' policy and 'pollution' policy, and greater attention being paid to the importance of products. In future, pollution control may come to be seen as less to do with emissions from stationary plant and more to do with how we use products in the home and on the farm, and the kinds of products that are allowed.

Paper delivered October 1997

Developments since 1997

All three approaches to integration (substance, source and region) described in the OECD guidance are now firmly part of EU legislation.

There have been amendments but the essential elements of the IPPC Directive 96/61 remain in place and embody the **source approach** – the sources being industrial installations broadly defined to include intensive pig and poultry production. After a long and wide-ranging review, the Directive was amended and consolidated with others – including the 'large combustion plant' Directive (see Chapter 4) – into the 'industrial emissions' Directive 2010/75.

Two institutional arrangements have evolved that should help the Directive to achieve its objectives.

- The European IPPC Bureau was established in Seville in 1997 as part of the European Commission's Joint Research Centre. It produces technical documents drawing on all available information and setting out the best available techniques (BAT) for controlling emissions from plants in different industrial sectors. These are called BREFs (BAT reference documents). They also set out the expected levels of emissions to air, water and land. More than one technology may embody BAT. BREFs are used as guidance by national authorities when authorising/permitting individual plants and setting emission limits. They also point industrialists to the best available technology. BREFs are nowhere mentioned in the original IPPC Directive 96/61, but the establishment of the IPPC Bureau was an appropriate administrative response to the requirements in the Directive for an exchange of information between Member States on their assessment of BAT and on emission limits. The 'industrial emissions' Directive 2010/75 now formally requires the Commission to produce and update BREFs – a task still carried out by the IPPC Bureau – and also to convene a regular forum of Member States, industries and NGOs to give views on the exchange of information and the quality of BREFs.

- IMPEL (now called the EU Network for the Implementation and Enforcement of Environmental Law) began life in 1992 as an informal network of authorities within the Member States responsible for industrial installations. In 2009 it became a legal entity. It receives financial support from the Commission and the Member States. It promotes the exchange of information, carries out joint enforcement projects, and encourages capacity building and training of inspectors. Based on IMPEL's work, the Commission issued Recommendation 2001/331 on 'minimum criteria for environmental inspections'. This recommended Member States to develop plans and programmes for inspections of industrial installations. The 'industrial emissions' Directive now makes these plans and programmes mandatory.

These two institutional arrangements should stimulate industrialists to adopt the cleanest technology and to use the least energy and resources. The use of inefficient 'end-of-pipe' solutions controlling discharges to just one medium should accordingly be discouraged. An equivalent of the IPPC Bureau is unlikely to have been established in all Member States if only for reasons of cost. It lifts that burden off their shoulders and ensures that information about the best technology is collected efficiently and spread throughout the EU. Since BREFs are published, they are available worldwide.

But has the Directive resulted in EU industry becoming cleaner in the nearly twenty years it has been in force? Curiously, this question was not asked during the review of the Directive, which concentrated on how it was working. We know from a report of 2009 that there were 43,264 installations in the EU (of twenty-seven Member States) covered by the Directive. We also know that industry prefers having only one integrated permit – a 'one-stop shop' – rather than dealing with separate authorities each only interested in discharges to one medium. Fortunately, there is a mechanism that could be used to answer the question. Directive 96/61 required a 'pollution emission register' inspired by something similar in the USA. Regulation 166/2006 now defines the information required of the 'European Pollutant Release and Transfer Register' (E-PRTR), and from that it should be possible to show if plants are becoming cleaner.

The **regional approach** has now also been firmly endorsed in the EU. As shown in Chapter 5, the river basin management plans required by the 'water framework' Directive 2000/60 have to identify all human pressures and must contain measures to be adopted to achieve 'good' water status. Leachate from land and pollutants deposited from the air and washed into rivers all have to be taken into account. The 'marine strategy framework' Directive 2008/56 requires similar plans for the four European seas into which all EU rivers discharge.

The Berlin paper above gave examples of how the **substance approach** was being practised in the 1990s. This continues, though more extensively because the REACH Regulation 1907/2006 requires the registration of many thousands of existing chemicals to be accompanied by a safety assessment. Each assessment has to cover potential effects on human health and on the environment including its aquatic, terrestrial and atmospheric 'compartments' (see Chapter 7). The information so generated can be used to restrict the marketing of chemicals. It also informs those designing products and those managing industrial installations and river basins. The source, substance and regional approaches can therefore all be linked.

The role of IEEP and the CF/IEEP project is described in the Appendix.

Notes

1 The Conservation Foundation was later to merge with the World Wildlife Fund (USA).
2 A 1984 CF report headed *State of the Environment: An Assessment at Mid-Decade* had a chapter on 'Controlling cross-media pollutants' with a section entitled 'A change in focus'. This noted that the US EPA was already testing three focuses.
3 Use of the word 'inspectorate' can cause confusion. In most countries an 'inspectorate' inspects, though it may also have powers of enforcement. In the UK, HMIP also authorised plants.

References

Davies, T (1990). The United States: Experiment and Fragmentation. In: Haigh, N and Irwin, F, eds, *Integrated Pollution Control in Europe and North America*. Washington, DC: Conservation Foundation, pp 51–66.

Emmott, N and Haigh, N (1996) Integrated pollution prevention and control: UK and EC approaches and possible next steps, *Journal of Environmental Law*, 8(2), pp 301–311.

Haigh, N (1998) Integrating environmental law – existing laws, aims, possibilities, *Environmental Law: Journal of the UK Environmental Law Association*, 12(2) (previously published in German by the German Council on Environmental Law).

Haigh, N and Irwin, F, eds (1990) *Integrated Pollution Control in Europe and North America*, Washington, DC: Conservation Foundation.

IEEP (1989) *Possibility for the Development of a Community Strategy on Integrated (Multi-Media) Pollution Control*, London: Institute for European Environmental Policy.

OECD (1991) *Integrated Pollution Prevention and Control*, Environment Monograph No. 37, Paris: OECD.

RCEP (1976) *Air Pollution Control: An Integrated Approach*, Royal Commission on Environmental Pollution 5th Report, London: HMSO.

9

CLIMATE CHANGE

We have seen how the EU acquired the power to enter into international agreements in Chapter 2, which also described the EU's first policy declaration on climate change in 1990. This was a call to stabilise or cap total EU emissions of CO_2 at 1990 levels by the year 2000, which then influenced the UN Framework Convention on Climate Change (FCCC). The book chapter reprinted below (in a much shortened form) tells the story of the beginnings of EU climate policy and the conflicts over the kind of EU legislation that was needed to turn the declaration into a form that enabled the EU to fulfil its obligations under the FCCC (Haigh 1996). The decision to cap emissions, followed by cuts, has provided the foundation of subsequent policy, but how the Member States share the burden (or the effort, as it is now called)[1] to achieve the obligations remains an issue. In microcosm, the EU's problems are those of the world. There is no doubt that at key moments the EU has played an influential role internationally, a role which no Member State on its own could have achieved.

The book in which the chapter appeared was the result of a collaborative project that described the evolution of climate policy in several countries and the EU, as well as the FCCC (O'Riordan and Jaeger 1996). Included in the book is a chapter on Germany, which was by far the largest emitter of CO_2 in the EU and which then played a major role in the EU's decision to cap emissions.

Climate change policies and politics in the European Community – abridged version of a book chapter published in 1996

This chapter places EC climate policy in the context of pre-existing EC policy for the environment and for energy. It concludes with observations on the importance for the success of the FCCC of a Community that actually meets its collective target, and provides a sound political and economic basis for further reductions of greenhouse gases beyond 2000.

EC policy-making

The EC 'legislature' is composed of the Commission, the Council and the Parliament acting together in accordance with the Treaty. The Commission may propose legislation and only the Council can finally adopt it. The Parliament, depending on the subject matter under discussion, has a greater or lesser influence, including in some circumstances the power of veto.[2] Where EC legislation can be adopted by qualified majority voting in the Council, a Member State may have to accept obligations to which its government is opposed. It is in these circumstances that the EC most clearly assumes a supranational character. Under such conditions the Parliament is also given a greater role than merely to give an opinion, thus providing some greater legitimacy to the ultimate decision to overrule a particular Member State. The Parliament's greater role is manifested in the 'cooperation procedure' or the 'co-decision procedure' which gives the Parliament certain extra powers to modify proposals and in the extreme to veto them (Wilkinson 1992). The Treaty of Maastricht extended the number of occasions when qualified majority voting (QMV) applies, so that it has become the standard procedure for environmental legislation. Nevertheless, the Treaty specifically applies exceptions for which unanimity is required. These are particularly relevant in the field of climate change since they include 'provisions primarily of a fiscal nature' and 'measures significantly affecting a Member State's choice between different energy sources and the general structure of its energy supply'. Despite this the EC adopted a decision to ratify the FCCC by qualified majority.

A major difficulty in discussing EC policy for climate change, or indeed for any other subject, is the ambiguous way in which the term EC policy is often used. As we have seen, the EC's tasks are carried out by a number of Institutions. The Institutions of the EC are not themselves the EC, and a proposal made by the Commission does not become EC policy until it has been adopted in a formal text by the Council following the proper procedures laid down in the Treaty. Similarly, an opinion of the Parliament is not EC policy, nor yet is a political declaration by the Council until it is embodied in a legally binding text. Such opinions or declarations may be precursors to EC policy and give an indication of possibilities. Since the Commission is the initiator of EC legislation in which policy is embodied, it is easy for the policies which it is advancing to be confused with EC policy. But strictly this is incorrect. Although the Commission put forward a proposal for a carbon and energy tax, such an idea would have only become EC policy if the Council had adopted a formal text on the subject, which it did not. The Commission has much less power over the Council than most national governments do over their own legislature, as is shown by the fact that many proposals from the Commission are never adopted. It is perfectly proper to talk of the 'policy of the Commission' once it has made a formal proposal but that does not mean it is, or will become, EC policy.

EC energy policy

The EC's involvement with energy is still not formally provided for in the core Treaty, although the associated Treaties that created the European Coal and Steel Community and the European Atomic Energy Community influenced the coal industry and many aspects of the nuclear industry since the origins of the EC.[3] Attempts to form an EC energy policy have had to confront the very different pattern of energy supply in Member States, some of which are richly endowed with energy resources while some have none – apart from the Sun. Some Member States depend heavily on nuclear power (France) while some are totally opposed to it (Denmark). Following the oil price rises of the 1970s, the EC began developing policies for energy efficiency in the interests of reducing energy imports, which is a common concern of all Member States. In 1986, for example, the EC adopted a target of improving the efficiency of final energy use by at least 20 per cent by 1995 for reasons unrelated to climate change. When in 1988 it became clear that this target would not be met without more vigorous action being undertaken a new range of measures was initiated which came to be associated with climate change policies. These include the SAVE and ALTENER programmes on energy efficiency and renewable energy, respectively, and a range of research and development programmes (notably JOULE and THERMIE).

EC environmental policy

It was not until the Single European Act amended the Treaty of Rome in 1987 that express authority for an environmental policy was provided. The Act established certain principles of environmental policy, including the principle that preventive action should be taken; that environmental damage should as a priority be rectified at source; that the polluter should pay; and that environmental protection requirements should be a component of the EC's other policies. This last principle is of very considerable importance for climate policy, and was strengthened by an amendment made by the Maastricht Treaty to become 'environmental protection requirements must be integrated into the definition and implementation of other Community policies'. The Maastricht Treaty added the precautionary principle which is also of considerable importance for EC climate policy (see Chapter 13).

The Single European Act also introduced the 'principle of subsidiarity' expressly for environmental policy and the Treaty of Maastricht then extended the principle to all EC policies (see Chapter 12). It confines action by the EC to those matters whose objectives cannot be sufficiently achieved by the Member States and that can be better achieved by the EC. This has had implications for EC climate policy as we shall see below.

One topic that was successfully handled by the EC, and that has provided some lessons for the climate change debates, is acid rain (see Chapter 4). In the mid-1980s when the issue was finally forced onto the agenda of reluctant countries, the

EC agreed on legislation which should result in an EC-wide reduction in sulphur dioxide emissions from large combustion plants of 58 per cent by the year 2003 compared with a 1980 baseline. This experience gave the Commission the confidence that it could tackle a major environmental issue transcending national frontiers with considerable implications for the cost of energy. A useful precursor to the subject of climate change was thus provided. Simultaneously in the mid-1980s the EC was involved in the negotiation of the Vienna Convention on the Protection of the Ozone Layer, and its associated Montreal Protocol, an involvement that not only influenced the final form of the Montreal Protocol, but also delivered intact a bloc of twelve countries as signatories to the Convention. This outcome would by no means have been a foregone conclusion without the EC (see Chapter 2).

By the end of the 1980s, therefore, when the issue of climate change reached the political agenda, the EC was in a position where it had acquired sufficient experience of environmental policy-making to feel able to take on a leading role.

The development of EC climate policy

EC policy for climate change can conveniently be regarded as having evolved in three phases which ended with ratification of the Convention by the EC in December 1993. We are now effectively in a fourth phase.

First Phase – 1988–90

It was the European Parliament, in a Resolution in 1986, that was the first of the EC Institutions to recognise the need for EC policy on climate change. This followed a scientific conference at Villach in 1985, which concluded that the time was ripe for policy initiatives on climate change.

It was not until November 1988, however, that the Commission issued the first of its communications on the subject [COM(88)656]. Its Fourth Action Programme on the Environment of 1987 had made no mention of climate change except as a subject for further research. The Commission's communication of November 1988 reviewed the scientific findings and possible actions to be taken without making any precise recommendations beyond proposing a work programme. In the same month the Intergovernmental Panel on Climate Change (IPCC) was established (see Chapter 10). The earliest EC developments can therefore be seen to have largely coincided with developments elsewhere.

The seriousness of the subject was confirmed at the highest EC level in June 1990 when the European Council (composed of heads of state and government) meeting in Dublin called for early adoption of targets and strategies for limiting emissions of greenhouse gases. This paved the way for political agreement, at a joint Council of Energy and Environment Ministers in October 1990, that CO_2 emissions should be stabilised in the EC as a whole by the year 2000 at 1990 levels. The wording of the Council's conclusions included the following:

> EC Member States and other industrialized countries should take urgent action to stabilize or reduce their CO_2 and other GHG emissions. Stabilization of CO_2 emissions should be in general achieved by the year 2000 at 1990 levels, although the Council notes that some Member countries according to their programmes are not in a position to commit themselves to this objective. In this context countries with, as yet, relatively low energy requirements, which can be expected to grow in step with their development, may need targets and strategies which can accommodate that development, while improving the energy efficiency of their economic activities. The European Community and Member States assume that other leading countries undertake commitments along the lines mentioned above and, acknowledging the targets identified by a number of Member States for stabilizing or reducing emissions by different dates, are willing to take actions aiming at reaching stabilization of the total CO_2 emissions by 2000 at 1990 levels in the Community as a whole.

This declaration was precipitated by the Second World Climate Conference in November 1990 and enabled the EC to take a strong and leading role, particularly in relation to the United States. This political agreement was not at that time embodied in any legally binding EC text and was in any event qualified by the assumption that other leading countries undertook commitments along similar lines. The qualification was not spelled out in any detail but has always been understood to include at least the United States, the world's largest emitter of CO_2.

While the political commitment to stabilise was a conscious attempt by the EC to show itself as a leader, it was also based on a certain pragmatism. Several Member States had by then adopted national policies to curb, stabilise or reduce greenhouse gas emissions. By making assumptions on what might happen in Member States that had adopted no national targets, it appeared that stabilisation was an attainable goal for the EC within ten years.

This first phase of EC policy-making on climate change, which can be said to extend from the recognition of the problem by the Commission to the adoption of a political target, i.e. the period from 1988 to 1990, developed remarkably rapidly despite inevitable resistance from some quarters. The political decision by the EC influenced the course of negotiations on the FCCC. Despite its non-binding character, Article 4(2) of the Convention would certainly have been much weaker without the EC's prior position. The Article was a compromise negotiated between the United States and EC Member States. The political weight of a combined EC put a reluctant United States under pressure.

Second Phase – 1990 to Rio

The second phase of EC policy-making can be said to span the period between the political agreement in 1990 and the signing of the Convention at Rio in June 1992.

The Commission, doubtless encouraged by the success of the first phase, made no secret of its desire to see the EC maintain its leadership role. It had been encouraged to do this by the declaration on the 'Environmental Imperative' issued by the European Council at Dublin in June 1990, which stated:

> The Community and its Member States have a special responsibility to encourage and participate in international action to combat global environmental problems. Their capacity to provide leadership in this field is enormous.
>
> *(EC 1990)*

Given this encouragement, it is no surprise that the Commission started by adopting an approach in which the major elements were to be decided at EC level. With hindsight it can be seen that this approach was overambitious and thereby delayed by perhaps a year the turning of the political commitment into a legally binding text.

Immediately after the joint Energy and Environment Council of October 1990, an internal draft of a Communication was prepared by the Directorates-General for Environment and for Energy. It stated that stabilising CO_2 emissions actually meant reducing emissions growth by some 10 to 20 per cent. Energy efficiency actions already underway, like the SAVE programme, seemed unlikely to secure such results. Accordingly, an adequate strategy was to consist of several measures complemented by economic and fiscal incentives. Action in the transport sector was also recognised as being necessary, including speed limits for cars and improved traffic and transport management. By May 1991 a further draft identified four major elements of climate policy, namely (a) a regulatory approach, (b) fiscal measures, (c) burden sharing among Member States, and (d) the scope for complementary action at the national level.

The concept of burden sharing, which the Netherlands in particular pressed for during its Presidency in the second half of 1991, was derived in part from the example of the EC's policy response to the issue of acid rain which it developed in the mid-1980s (see Chapter 4). Under the 'large combustion plants' Directive of 1988 Member States were allocated different targets for reducing emissions of sulphur dioxide from their existing power plants. The experience of the long drawn-out negotiations on this Directive should have given some warning of the difficulties inherent in this approach which could better be described as horsetrading rather than the sharing out of a burden based on objective criteria. The idea of a Directive allocating different targets for CO_2 emissions was eventually abandoned, and fiscal measures, i.e. a carbon energy tax, then became the cornerstone of the Commission's proposals. One objection, not often publicly mentioned, to the idea of national targets for CO_2 emissions being set at EC level was the transfer of powers that this would have implied over a matter which touched so many aspects of national life. Some Member States were reluctant to grant the EC so much competence, an issue discussed more fully below.

The deadline that came to concentrate the minds of the Commission and the Member States was the United Nations Conference on Environment and Development (UNCED) to be held in Rio in June 1992. If the EC was to maintain its leadership role which it had forged for itself at the Second World Climate

Conference in November 1990, it was necessary to have some policy flesh to put on the bones of the political decision to stabilise CO_2 emissions. It would be necessary to demonstrate how the EC was to achieve this. One course would have been to translate the decision on stabilisation into a legally binding instrument. This could also have required Member States to develop their own programmes with their own targets. Such an instrument could have been agreed fairly quickly after October 1990. Not surprisingly, given the Commission's traditional desire to expand EU competence, it chose the path of trying to develop a complete and ambitious package of measures to be agreed together. The most ambitious of these, the carbon energy tax, proved to be the most contentious, and delayed all the others. As finally announced in June 1992, these measures consisted of four elements, described more fully below:

- a framework Directive on energy efficiency within the SAVE programme
- a Decision on renewable energy – ALTENER programme
- a Directive on a combined carbon and energy tax, and
- a Decision concerning a monitoring mechanism for CO_2 emissions.

Third phase – Rio to ratification

The third phase spans the period from the signing of the Convention at Rio in June 1992 to its ratification by the EC in December 1993. During this period the Council had to decide whether to accept the four detailed proposals that the Commission had put forward just before Rio, the most important of which in the eyes of the Commission was the proposed carbon energy tax.

The carbon energy tax

The evolution of the tax idea has been described elsewhere (Liberatore 1995: 61–69). Although the Fourth Environmental Action Programme of 1987 proposed no specific measures, evidence on global warming had begun to be collected under a Commission-initiated research programme from 1979 to 1986. Proposed measures were then developed through an inter-service group of officials representing the Directorates-General on tax, competition, environment, energy, regional policy and research. The individual responsible for promoting the international leadership role for the Commission was its Commissioner for the Environment, Carlo Ripa di Meana. He promoted the cause of the tax in meetings in Washington in 1990, and at the pre-Rio conference on sustainable development held at Bergen in the same year.

Three elements combined to advance discussions, namely the inter-service group initiative; a belief in the need to show leadership (urged on the Commission by the Germans and the Dutch in particular); and a joint meeting of energy and environment Ministers. The key to innovative reform lay in the connection between energy futures and climate change on the one hand, and economic development and energy efficiency measures on the other. The economic gains of energy conservation stimulated the belief that the tax could be a valuable trigger in economic renaissance. In this endeavour, the European Parliament also became involved, holding hearings and initiating debate. It is of interest to note that the inter-service group largely left out transport and agriculture in its deliberations.

The proposed tax was to be shared equally between carbon content and energy content on the grounds that a pure carbon tax would have favoured nuclear energy. For this reason alone the joint carbon and energy tax was always opposed by France, which favoured a straight carbon tax. The tax was to be progressive starting at the equivalent of 3 US dollars per barrel of oil and rising by 1 dollar per barrel per year until 10 dollars had been reached. The tax had opponents within the Commission, within industry and among the Member States. Agreement within the Commission to the proposal was only secured by exempting energy-intensive industries and making the proposal conditional on comparable action by other OECD countries. For a time there were some hopes that the United States might introduce an energy tax at a level that could be said to be comparable. When that hope was not realised, it was clear that either the conditionality element would have to be dropped or the tax would have to be abandoned. In the Council the tax was opposed by France, by the less developed Member States who argued that their development would be impeded, and by the UK. The UK's objection was not to the idea of fiscal measures but to the principle that taxation is a matter that should be the responsibility of the Member States. Since, even under the revised rules for decision making in the Council, fiscal matters can only be decided by unanimity, it was improbable that an EC carbon energy tax would be agreed. In December 1994 the Council eventually determined that there would be no tax set at EC level, but Member States were encouraged to develop their own taxes. This was regarded as a severe setback by the 'northern' Member States (Germany, the Netherlands, Denmark, Sweden, Finland), and the idea certainly did not go away.

Renewable energy

The Decision on renewable energy (ALTENER) is a programme of EU funding amounting to about 35 million ECU over five years. This is estimated to achieve a reduction of CO_2 emissions of 180 million tonnes by 2005. The agreed funding was smaller than originally envisaged by the Commission but otherwise proved uncontentious and has been adopted. Whether it will meet the target set for it is a moot point.

Energy conservation

The Directive on energy saving measures (SAVE), as adopted, places an obligation on Member States to introduce national programmes relating to a number of matters. On the grounds of subsidiarity it was much weakened from the original proposal made by the Commission. This is in addition to the laying down of EU standards for traded products such as central heating boilers and the requirement for energy labelling of products.

The monitoring mechanism

The last of the four elements is much more important than its title might imply. Not only is the Commission to evaluate data on greenhouse gases reported by the Member States but, as described more fully below, the Member States are to 'devise, publish and implement national programmes for limiting anthropogenic emissions of CO_2'. Because the Decision is more precise about the nature of the national programmes than is the Convention, and since the EC target is also sharper than in the Convention, the Decision must now be regarded as the cornerstone of EC climate policy.

If the EC carbon energy tax, which was seen by many in the Commission as the cornerstone, has been abandoned at least for the time being, then the monitoring mechanism forces responsibility back on the Member States, not only to set their own targets but to devise national measures to ensure that national targets are achieved. An important role for the Commission is to ensure that the national programmes add up to the EC stabilisation target, and to sound the alarm if they do not. One of the regrettable features of the saga of the proposed carbon energy tax is that it delayed the monitoring mechanism by about a year. As it is, the national programmes so far submitted to the Commission are inadequate and do not yet enable firm conclusions to be drawn as to whether the stabilisation target will be met.

Ratification

A curious side-show in the process leading to ratification of the Convention was an attempt by six countries, led by the Netherlands and Germany, to link ratification to adoption of the EC carbon energy tax. The UK called this bluff by refusing to change its view on an EC tax, and by implying that it would not be particularly concerned if the EC did not ratify the Convention and that it would ratify on its own. This it did, quickly followed by Germany and then the others. The idea of the EC not ratifying caused concern to those countries like Spain that were intending to offset their planned increase in CO_2 emissions against reductions by other countries (e.g. Germany, the Netherlands). In the event a face-saving formula was found which enabled the EC to decide to ratify. The formula left the decision on the tax to the Council of Finance Ministers (with whom the decision rested anyway) and emphasised the possibility of national taxes which might amount to something equivalent to the proposed EC tax. Had the Netherlands not withdrawn its opposition to EC ratification, then it is possible that some Member States, such as Portugal and Spain, would not have ratified on the grounds that on their own their national plans could not be said to live up to the requirements of the Convention. In the event all Member States, as well as the EC, ratified the

Convention. The EC will accordingly have to submit a report to a future Conference of the Parties on how it will fulfil the obligations in the Convention.

Fourth phase: follow-through

We are now in a fourth phase where the attention is on how far the national programmes will achieve the stabilisation target.

The 'monitoring mechanism'

The monitoring mechanism was adopted by the Council as Decision 93/389 in June 1993. Although the Decision does not expressly say so, the mechanism consists of the Commission consolidating and evaluating certain information submitted by the Member States. The Decision also requires Member States to 'devise, publish and implement national programmes for limiting their anthropogenic emissions of CO_2' in order to contribute to (a) the qualified target for stabilisation of CO_2 emissions by 2000 at 1990 levels in the EC as a whole (the qualification being the assumption that other leading countries undertake commitments along similar lines); and (b) the fulfilment of the commitment relating to CO_2 in the Convention by the EC as a whole. The programmes are to be updated periodically with a frequency to be decided by the Commission. The contents of the programme are specified more precisely than the equivalent requirements in the Convention, and from the first update each national programme is to include:

- the 1990 base year anthropogenic emissions of CO_2;
- inventories of national anthropogenic CO_2 emissions by sources and removal by sinks;
- details of national policies and measures limiting CO_2 emissions;
- trajectories for national CO_2 emissions between 1994 and 2000;
- measures being taken or envisaged for the implementation of relevant EC legislation and policies;
- a description of policies and measures in order to increase sequestration of CO_2 emissions;
- an assessment of the economic impact of the above measures.

One important difference from the Convention is the additional requirement to provide 'Trajectories of national CO_2 emissions between 1994 and 2000'.

The Decision requires Member States annually to report to the Commission on their CO_2 emissions and removal by sinks for the previous calendar year, which again is a more precise obligation than in the Convention. Within three months of receiving the information, the Commission is to produce inventories for the whole Community and is to circulate them to all Member States.

The Decision also requires the Commission to evaluate the national programmes in order to assess whether progress in the Community is sufficient to ensure fulfilment of the commitments. Annually after the first evaluation, the Commission, in consultation with the Member States, is to assess progress and is to report to the Council and the Parliament.

National programmes

The first evaluation of the existing national programmes under the monitoring mechanism was issued by the Commission in March 1994 [COM(94)67]. This evaluation was based on the national programmes available to the Commission in early August 1993, not all of which were then complete. The national reduction goals set out in the programmes are shown in Table 9.1.

A comment is necessary about the difference between the word 'stabilisation' and the words 'return to'. 'Stabilisation' implies that emissions will not rise above the given figure after the year 2000, whereas the words 'return to' carry no such implication.

All countries except Germany have given CO_2 emission figures for the year 2000. Germany remains committed to its national target of reducing by 25 per cent by 2005 from a baseline of 1987. The exact amount of Germany's reduction between 1990 and 2000 has not been projected by the German Government, but Germany's CO_2 emissions decreased by 14 per cent between 1987 and 1993, although much of this reduction was due to the economic restructuring of the former East Germany. Some of this reduction was already achieved before 1990 so that Germany's possible total reductions between 1990 and 2000 remain problematic.

From a knowledge of national emissions in 1990 it is possible to calculate what the emissions might be in the year 2000 (Table 9.2) assuming that national goals are fulfilled.

TABLE 9.1 CO_2 reduction goals in national programmes in 1993

Country	*Percentage*	*Detail*
Belgium	5	Reduction from 1990 by 2000
Denmark	5	Reduction from 1990 by 2000
France	13	Increase from 1990 by 2000
Germany	25–30	Reduction from 1987 by 2005
Greece	25	Increase from 1990 by 2000
Ireland	20	Increase from 1990 by 2000
Italy	Stabilisation on 1990 level by 2000	
Luxembourg	Stabilisation on 1990 level by 2000	
The Netherlands	3–5	Reduction from 1989/90 by 2000
Portugal	30–40	Increase from 1990 by 2000
Spain	25	Increase from 1990 by 2000
UK	Return to 1990 level by 2000	

TABLE 9.2 National CO_2 emissions in million metric tonnes

Country	*1990*	*2000*
Belgium	112.0	106.4
Denmark	53.01	50.36
France	365.7	413.24
Germany	1,005.0	1,005.0 (assuming no change)
Greece	73.7	92.13
Ireland	30.8	36.96
Italy	402.4	402.4
Luxembourg	12.5	12.5
The Netherlands	157.3	152.58 (assuming –3%)
Portugal	39.9	55.86 (assuming +40%)
Spain	210.7	263.38
UK	579.2	579.2
Total EC (twelve Member States)	3,042.21	3,170.01= 4% increase on 1990

Since Germany is by far the largest emitter of CO_2, responsible for nearly a third of all EC emissions in 1990, the assumption made about German emissions in 2000 will dramatically affect the total EC result. On the conservative assumption made in Table 9.2 that Germany stabilises its emissions by 2000 at 1990 levels, it can be calculated that the EU as a whole will show a 4 per cent increase in 2000 compared with 1990. If the assumption is made that Germany reduces its emissions in 2000 by 12.5 per cent compared with 1990 (i.e. half the 25 per cent reduction proposed by 2005 compared with 1987), its CO_2 emissions in 2000 become 879.4 metric tonnes, with the result that total EC emissions in 2000 are approximately equal to the 1990 total. The EC stabilisation goal would then be achieved, assuming that the other national goals are achieved.

The Commission will come forward with further evaluations based on revised national programmes and inventories of emissions and as time goes by a more accurate assessment of what might happen in 2000 will emerge. But from the evidence so far, several countries, including the four so-called 'cohesion countries', are assuming that they will increase their emissions considerably. France is projecting an increase of 13 per cent because it is assuming increased consumption of fuel in the transport sector, with little scope for compensating reductions in emissions from electricity production since it is largely derived from nuclear power stations. The cohesion countries all project increases of 20 per cent or more. This explains why they, and particularly the largest of them, Spain, were insistent that the EC should ratify the Convention since if their increases are subsumed in the EC total, and the EC as a whole achieves stabilisation, then their own increases can be presented as part of a form of 'joint implementation' by the EC as a whole. On their own they felt that they would not be able to fulfil the requirements in the Convention to endeavour to return emissions by 2000 to 1990 levels.

The cohesion countries

Spain, Portugal, Greece and Ireland all have a national GDP which is below 90 per cent of the EC average and are therefore eligible for payments from the Cohesion Fund. This is intended to strengthen economic and social cohesion within the EC through the provision of finance for projects relating to environmental protection and transport infrastructure. The group of four countries can be regarded as a distinctive category for climate change purposes for the following reasons:

- their per capita CO_2 emissions are relatively low;
- their combined contribution to the EC total of CO_2 emissions is fairly small (about 12 per cent);
- there is a presumption that their relatively low GDP will increase relatively rapidly to achieve the goal of economic convergence;
- as a result of the above, and based on the principle of burden sharing, there is an assumption that they will not be constrained to CO_2 targets as demanding as those of other Member States.

The cohesion countries present in microcosm the dilemma of selecting appropriate targets for the developing countries under the Convention. Relative to the richer EC Member States the cohesion countries have low per capita CO_2 emissions, but in a world context they are developed countries with relatively high per capita CO_2 emissions. If the EC as a whole accepts that joint implementation can entail the cohesion countries having unrestrained or only moderately restrained increases in CO_2 emissions, it is difficult not to follow the same argument for developing countries. A difference of course is that the cohesion countries represent a small proportion of the total EC population.

Burden sharing

Although the phrase 'burden sharing' was not used in the conclusions of the Council in October 1990 which first set out the EC's stabilisation target, the concept was there, and the phrase is part of the language in which EC climate policy is discussed. The national targets for curbing CO_2 emissions constitute a form of burden sharing that has been called the 'bottom-up' approach since the choice of national target has been made by the country itself. If the national plans of the Member States happen to result in the EC's stabilisation target being achieved, then the 'bottom-up' approach creates no problems. But this will not be so if it becomes clear that EC stabilisation will not be achieved. In 1991 the Commission had indeed begun work on a draft Directive that would have attempted to allocate

CO_2 emission reductions among the Member States on the model of Directive 88/609 concerned with emissions of sulphur dioxide from large combustion plants. This approach met opposition and was abandoned. The experience of the large combustion plants Directive suggests the difficulties entailed in trying to set legally binding targets for a number of countries each having quite different circumstances. Even if the 'top-down' approach has been abandoned for the time being, it remains implicit in the setting of an EC target (i.e. stabilisation) since, if it looks as though the target will not be achieved by the 'bottom-up' approach, some process of negotiation between the Member States and the Commission will have to take place in which pressure will be applied on those Member States that are perceived by the others as not having adequately shouldered their share of the EC burden. A version of the 'top-down' approach may well therefore re-emerge but possibly in an informal way. The negotiations may need to be based on some attempt at an objective evaluation of what fair burden sharing should entail.

Competence

A particular reason why some Member States can be expected to resist a 'top-down' approach to burden sharing, although this may not be stated publicly, is the loss of sovereignty entailed. Even if it is possible to agree a fair allocation of reductions among the Member States, the setting of these percentage figures in a Directive has considerable implications for the future, because competence is transferred to the EC from the Member States for the setting of further targets (see Chapter 2).

Qualified majority voting

The question of loss of national sovereignty is a concern of all Member States though some express the concern more openly than others. It is a concern even when decisions are made unanimously in the Council. But the concern becomes much more acute once there is the possibility of a Member State being outvoted in the Council and having to accept some obligation, such as an emission target, against its wishes. Before the Single European Act, all EC environmental legislation had to be adopted unanimously. The Single European Act introduced qualified majority voting (QMV) for legislation affecting the internal market, and the Treaty of Maastricht made QMV the normal method of adopting all environmental legislation, albeit with some exceptions. These exceptions are fundamental to climate policy because they include 'provisions primarily of a fiscal nature', which therefore covers the proposed carbon energy tax, and also 'measures significantly affecting a Member State's choice between different energy sources and the general structure of its energy supply', which must include many measures to curb CO_2 emissions. However, measures dealing with the energy efficiency of traded products such as gas boilers can certainly be adopted by QMV under the Article in the Treaty concerned with the single market, and therefore already some climate change-related measures are subject to QMV. The introduction into the Maastricht Treaty of the exceptions relating to 'measures significantly affecting a Member State's choice between different energy sources' shows how sensitive at least some Member States are to the prospect of being outvoted on such issues.

An indication of possible conflicts ahead is provided by the adoption of Council Decision 94/69 concerning the conclusion of the Convention during which the UK abstained. The Council legal services had advised that the Decision should be adopted by QMV while the UK argued that the Convention significantly affects a Member State's choice between different energy sources and that a Decision to ratify the Convention must therefore do the same. The UK's abstention appears to be in the nature of a marker for the future and the UK is unlikely to seek annulment of the Decision by the Court.

Subsidiarity

The principle of subsidiarity set out in the Treaty (see Chapter 12) does not deal with whether action should be taken within the Member State by the central government or by local or regional governments, since that is a matter for the Member States themselves. Nor does it deal with the question of whether action is better taken by international machinery, such as that created under the FCCC, rather than by the EC. Any broad view of subsidiarity must deal with these questions. Such a view was contained in one of the principles set out in the First Action Programme on the Environment of 1973 which states that 'in each category of pollution, it is necessary to establish the level of action (local, regional, national, Community, international) best suited to the type of pollution and to the geographical zone to be protected'.

The following two types of question therefore arise in the context of climate change. If the machinery of the Convention, to which all Member States are Parties, is adequate to deal with the climate issue, what is the justification for the EC duplicating that machinery? Similarly, if national (or local) action on some matters, such as the insulation of housing, adequately deals with that matter, what is the need for the EC to set insulation standards? The issue of subsidiarity is therefore very relevant for climate change policy and has indeed influenced decisions. Opposition to the carbon energy tax by the UK is not so much an objection to the idea of such taxes but to them being set at EC level rather than at national level. Equally, the SAVE Directive was amended in the Council by the removal of several detailed requirements such as EC-wide insulation standards for each climatic zone. Instead the SAVE Directive merely requires Member States to establish national programmes for a number of fields including the thermal insulation of buildings. The discussion about subsidiarity will therefore certainly continue in the field of climate change as it does in other areas, and the resolution of any particular conflict is likely to be pragmatic.

Policy integration

None of the four measures proposed by the Commission just before the Rio conference in 1992 is specifically directed at particular policy areas such as transport, industry or agriculture, although all are affected. The Treaty now requires that environmental protection requirements must be integrated into the definition and implementation of other Community policies which means that all EC policies must now contribute to climate change policies.[4] The principal EC document that sets out the framework for doing this is the Fifth Action Programme on the Environment, which is very different from the earlier programmes (see Chapter 1). It identifies actors, target sectors and key environmental issues. The first of the environmental issues listed is climate change. The five target sectors chosen are manufacturing industry, energy, transport, agriculture and tourism.

The Programme suggests a broader range of policy instruments than reliance on legislation, such as economic and fiscal instruments and voluntary agreements, and also sets out two guiding principles, namely the precautionary approach and the concept of shared responsibility. Shared responsibility is interpreted as meaning that all the major economic and political players in society have a role in implementing the Programme, including the general public both as citizens and consumers. The Programme is a wide-ranging document which sets ambitious targets intended to steer the EC on the path towards sustainable development. One result is that the Commission has adopted internal procedures which include the appointment of senior officials with special environmental responsibilities within each Directorate-General.

The way the mechanisms will work in practice remains unclear, but the political conflicts involved are considerable, particularly in the transport sector. Here the EC is promoting a trans-European road network which is bound to result in increasing road traffic and hence increasing CO_2 emissions. This conflict mirrors comparable conflicts within each Member State and its resolution will be long drawn out. What can be said at present is that climate change arguments reinforce at a strategic level other environmental arguments about road traffic which may have greater force at local level, such as air pollution or destruction of habitats. For that reason, they can be applied in strategic discussions such as those concerned with trans-European road networks. In the absence of an EC climate policy it would be much more difficult to bring environmental arguments to bear on EC level discussions about road building. The same will hold for other policy areas.

EC involvement in the Climate Change Convention

We have seen how the Council's political declaration to stabilise CO_2 emissions in the EC as a whole by the year 2000 at 1990 levels gave an impulse that led two years later to the signing of the Convention with stronger provisions than might otherwise have been expected. Had no Convention been agreed the EC would have then had to consider whether on its own it should have pursued a policy for climate change. A powerful argument would have been advanced that it should do so as an encouragement for the rest of the world. Once the Convention was signed the argument was different. One view is that action by the EC was no longer necessary, and that there was no real need for the EC to be a party to that Convention. Clearly the cohesion states did not share this view. They have had a very good reason for wanting the EC to be a party since otherwise it is doubtful that they fulfil the requirements of the Convention, given that they are projecting significant increases in CO_2 emissions by the year 2000. The 'cohesion countries' are relying on the EC as a whole to stabilise its emissions so that they do not have to.

The EC (as opposed to its Member States) in fact played only a limited role in the negotiations leading to the Climate Change Convention. One key issue was whether the EC should insist on the Convention including a firm commitment of 'stabilising' CO_2 emissions by the year 2000 at 1990 levels. The United States was firmly opposed to such a commitment and the EC was thus faced with two possibilities: to press for a strong Convention in the knowledge that the United States would probably not then sign, or to agree to a form of words which would enable the United States to sign while still taking account of the EC's own political commitment. There were some Member States which would have accepted a stronger Convention without US participation, while others argued that without US participation the Convention would hardly be worthwhile, and other countries would be given an excuse not to sign.

During these discussions the UK Secretary of State for the Environment, Michael Howard, allegedly with the encouragement of some other Environment Ministers from EC Member States, travelled to the United States and agreed a form of words with US officials which forms the basis of Article 4(2) of the Convention. Its convoluted language commits developed countries to modify their trends in emissions and effectively to endeavour to return by the end of the decade to earlier levels of emissions. Whether this can be regarded as an EC contribution to the framing of the Convention is a matter of opinion. Formally it was not, since no formal Council decisions were taken on the subject, but without the machinery provided by the EC for discussion between Ministers it might never have happened.

Before the first Conference of the Parties (COP) in March 1995, the EC urged it to establish a Protocol for setting targets and timetables to reduce greenhouse gases beyond 2000. The proposed dates were 2005 and 2010, though no figures were provided.

The ability of the EC to contribute to the evolution of the Convention at future COPs will depend on its ability to meet its own targets. If it cannot do that, the confidence with which it can take on a leading role will be much diminished.

Book chapter drafted December 1994 – published 1996

Developments since 1996

The monitoring mechanism has been revised to take account of the Kyoto Protocol and is now set out in Decision 280/2004. The progress report of 2002 [COM(2002)702] confirmed that the EU had met its target under the FCCC of stabilising emissions at 1990 levels by 2000 and was 3.5 per cent below it. This was largely as a result of one-off occurrences in two Member States: the collapse of manufacturing industry in East Germany following reunification, and the conversion of power generation in the UK from coal to gas.

Having urged the 1995 COP to set targets beyond 2000, the EU in 1997 called for a 15 per cent cut by developing countries of emissions of CO_2, methane and nitrous oxide by 2010. This target was supported by the G77 group of developing countries, and as it was conditional on other industrialised countries making their own contributions, the EU was able to apply pressure on the USA (the largest emitter) and Japan (the host of the COP held at Kyoto). After difficult negotiations the Kyoto Protocol to the FCCC was agreed with an overall reduction of greenhouse gases (including, additionally, hydrofluorocarbons, perfluorocarbons and sulphur hexafluoride) of 5.2 per cent by 2008–12, with the EU contributing 8 per cent, the USA 7 per cent and Japan 6 per cent. No targets were set for developing countries.

An analysis of the EU's role in the Kyoto Protocol and how its policies developed subsequently would require a book, and at least one exists (Jordan et al. 2010). That almost half the authors are also editors suggests that no single person can now follow the whole of this growing subject. This chapter can only point to key developments.

In 1998 the Council reached a 'burden-sharing' political agreement on how the EU was to achieve its collective Kyoto target of an 8 per cent reduction in greenhouse gas emissions. These are shown in Table 9.3.

To ratify the Kyoto Protocol, the Council adopted Decision 2002/358, which included the burden-sharing agreement. This became binding on the Member States once the Protocol entered into force. Although the USA had signed the Protocol, it then refused to ratify it. The Protocol could come into force only if enough other major polluting countries ratified, and this was achieved in 2005

TABLE 9.3 'Burden-sharing' agreement to meet EU Kyoto Protocol target of 8 per cent

Member State	*Reduction commitments (%)*
Luxembourg	–28
Denmark	–21
Germany	–21
Austria	–13
United Kingdom	–12.5
Belgium	–7.5
Italy	–6.5
Netherlands	–6
France	0
Finland	0
Sweden	+4
Ireland	+13
Spain	+15
Greece	+25
Portugal	+27

when Russia was persuaded to do so by the EU in exchange for supporting its application to join the World Trade Organization.

The European Climate Change Programme (ECCP) was launched in 2000 [COM(2000)88] by the Commission at the same time as a Green Paper on an EU emissions trading scheme. The first report of the ECCP set out a number of proposed policies and measures with an estimated cost-effective reduction potential of more than twice what was required to achieve the EU's Kyoto target of 8 per cent. Over a third of these were in the energy supply sector, while 40 per cent of the remainder would result from energy efficiency measures relating to buildings, products and industrial processes. The policies and measures proposed included some already under development (such as energy performance of buildings, and combined heat and power). A second report in 2003 outlined some progress but noted that the transport sector was particularly difficult. In 2005 a new round of the ECCP reviewed the policies already proposed and discussed prospects for including aviation in emissions trading, CO_2 capture and storage, reducing vehicle emissions, and adaptation to climate change.

The Green Paper [COM(2000)87] on emissions trading resulted in Directive 2003/87, intended to help meet the EU's 8 per cent Kyoto target. The scheme, known as the EU-ETS (emissions trading system) came into operation in January 2005. It requires the operators of certain plants (power stations, refineries and large, energy-intensive factories) to hold and surrender emission allowances to cover their carbon emissions. Allowances are either allocated freely to operators, or can be purchased from other operators or at auction. Firms that wish to increase their emissions must buy permits from those who have cut their emissions and are willing to sell their allowances. Similar schemes, sometimes known as 'cap and trade', operate elsewhere in the world. Such schemes depend on the existence of a cap on emissions, the theory being that the ability to trade in permits ensures that cuts in emissions are made where it is most economical to do so.

Arguably, the ETS has set out what it was designed to do. While there are some concerns over the treatment of biomass, and over the use of emissions reductions in developing countries to meet targets, emissions in the sectors concerned have broadly been kept within the cap. However, the introduction of the EU-ETS as a mechanism has not done away with the need for political will to set a sufficiently tight cap on emissions, and to adjust the cap downwards when necessary to reflect the reduction in emissions from an economic downturn. The carbon price created has been too weak to drive significant investment in the low-carbon alternatives required.

In 2007 the Commission's communication *Limiting Global Climate Change to 2°C: The Way Ahead for 2020 and Beyond* [COM(2007)2], proposed that the EU commit to a 20 per cent reduction by 2020, rising to 30 per cent if other countries take appropriate action too. This was supported by the Council and led to formal proposals from the Commission in 2008 known as the CARE (Climate Action and Renewable Energy) package, leading to the adoption of the so-called '20–20–20' targets: a 20 per cent reduction in EU greenhouse gas emissions from 1990 levels; an increase in the share of EU energy consumption produced from renewable resources to 20 per cent; and a 20 per cent improvement in the EU's energy efficiency below projected levels. The incorporation of those targets into Member State reporting requirements under the so-called 'European Semester' process, which is aimed at ensuring delivery of the key monetary and economic goals of the EU, is a welcome sign of the journey of climate policy to centre stage, although it seems clear that the attention paid to the climate and energy elements of the Semester process by both the Commission and Member States is relatively low.

One element of the CARE package was the 'effort-sharing' Decision 406/2009, which set the mandatory targets shown in Table 9.4 for each Member State for the

TABLE 9.4 Greenhouse gas emission limits for the non-ETS sectors by 2020 compared with 2005

Member State	*Percentage change from 2005*
Denmark	–20
Ireland	–20
Luxembourg	–20
Sweden	–17
Netherlands	–16
Austria	–16
Finland	–16
United Kingdom	–16
Belgium	–15
Germany	–14
France	–14
Italy	–13
Spain	–10
Cyprus	–5
Greece	–4
Portugal	1
Slovenia	4
Malta	5
Czech Republic	9
Hungary	10
Estonia	11
Slovakia	13
Poland	14
Lithuania	15
Latvia	17
Romania	19
Bulgaria	20
EU total	**–10**

non-ETS sectors. It was estimated that the ETS sectors emitted 60 per cent of the total and the non-ETS sectors 40 per cent. National targets were to be reached by 2020 as compared with 2005, the EU target being 10 per cent. The 2005 baseline was chosen because it was the first year that reliable data was available for both the ETS and non-ETS sectors. Progress is reported annually to the Commission under the 'monitoring mechanism' Decision 280/2004.

While the 'effort-sharing' Decision is a further step in EU competence on greenhouse gas emissions, the fact that the environment treaty base allows Member States to go further, if they choose, in limiting their emissions means that there is still some residual Member State competence.

The Kyoto Protocol foresaw negotiations starting on post-2012 commitments in 2005 and accordingly the Commission issued a Communication called *Winning the Battle against Global Climate Change* [COM(2005)35]. The European Council in 2005 committed the EU to pursuing policies designed to avoid global warming in excess of 2°C. The Portuguese Council Presidency in 2007 released a position paper calling on developed countries to reduce emissions by 60–80 per cent by 2050 compared with 1990 levels.

The COP held at Copenhagen in 2009, however, failed to show the degree of willingness to reach an agreement that had been evident at Kyoto. The Copenhagen Accord fell far below what was needed to tackle climate change, and effectively deferred real decisions on the emissions reductions that would be needed. Moreover, the final stages of negotiation not only departed significantly from the roadmap agreed at the Bali COP two years previously, but seemed to involve EU negotiators being bypassed while the USA and key emerging economies hammered out a deal. EU ambition and commitments on climate change seemed increasingly to be taken for granted by other negotiators.

In 2011 the Commission issued its *Roadmap for Moving to a Competitive Low Carbon Economy in 2050* [COM(2011)112] setting out milestones to achieve an 80 per cent reduction by 2050 and providing a strong case for a 25 per cent reduction by 2020. This was in response to the European Council's confirmation of the need for an 80–95 per cent cut in order to stay below 2°C (see Chapter 15). The increase in ambition at European level, from the hesitancy expressed when initial discussions took place on burden-sharing in the 1990s, is striking. One can speculate on a range of explanations for it, perhaps the most prominent of which are:

- a growing political understanding of the scientific consensus, and of the challenges of global action – it is simply not plausible for even reluctant leaders to claim that the problem does not need to be addressed, or can be postponed;
- experience with climate policy thus far has suggested that progress can be made without imposing excessive costs, and even with countervailing economic benefits, including the emergence of new sectors such as renewable energy;
- the availability of additional revenues, for example through auctioning of ETS allowances, creates new opportunities for transferring funds from wealthier Member States to finance low-carbon investment;

- a mainstreaming of popular support, reflected in the fact that the key groups in the European Parliament, to left, right and centre, are in principle in favour of action, even where there are some concerns among individual national delegations;
- some of the large Member States normally associated with concerns over the balance between environmental regulation and the interests of industry, particularly the UK and Germany, have been among the more positive on the need for action.

Recent experience suggests that even this level of broad support does not guarantee the adoption of targets in line with long-term climate mitigation goals; but it is worth noting that this level of mainstream agreement on an environmental objective is unusual. It will be important to maintain it and deepen it.

As this book goes to press, preparations are in hand for the Paris COP in December 2015. There are welcome signs of a greater political willingness, particularly from the USA and China, to introduce effective policies aimed at controlling and reducing emissions; and signs, too, that those policies will be reflected in international commitments. The willingness to reach a deal will, it is to be hoped, be significantly greater than at Copenhagen. However, the looser nature of 'Nationally Determined Contributions', and the prospect that they will not be enforceable under international law, while they contribute to the chances of a deal, raise real questions about the effectiveness of any deal in limiting global emissions sufficiently. Differing views among negotiators on the importance of legal enforceability, and the differing nature of commitments proposed by the parties, leave an enormous level of detail still to be resolved in a short period. However, it seems clear that there will be some continuing international framework for emissions reduction, with the EU aiming to play a leading role in demonstrating what can be achieved; and that role will continue to have important implications for EU internal policy-making.

Notes

1. The Commission has decided that the words 'effort sharing' sound more positive than 'burden sharing'.
2. In the early 1990s the Parliament had less power than it does today.
3. Energy policy is now the subject of Article 194 TFEU (Treaty on the Functioning of the EU). The Coal and Steel Community ceased to exist in 2002.
4. An example is the 'landfill' Directive 1999/31, the main purpose of which is to reduce emissions of methane – a powerful greenhouse gas – from biodegradable waste by reducing the amount of such waste in landfills (see Chapter 6).

References

EC (1990) Declaration by the European Council on the environmental imperative, *Bulletin of the EC*, 23(6), pp 16–18.

Haigh, N (1996) Climate change policies and politics in the European Community. In: O'Riordan, T and Jaeger, J, eds, *Politics of Climate Change: A European Perspective*, London: Routledge.

Jordan, A, Huitma, D, van Asselt, H, Rayner, T and Berkhout, F, eds (2010) *Climate Change Policy in the European Union*, Cambridge: Cambridge University Press.

O'Riordan, T and Jaeger, J, eds (1996) *Politics of Climate Change: A European Perspective*, London: Routledge.

Liberatore, A (1995) Arguments, assumptions and the choice of policy instruments: the case of the debate on the CO_2 energy tax in the European Community. In: Dente, B, ed., *Environmental Policy in Search of New Instruments*, Dordrecht: Kluwer, pp 55–72.

Wilkinson, D (1992) Maastricht and the environment: the implications for the EC's environmental policy of the Treaty on European Union, *Journal of Environmental Law* 4(2), pp 221–239.

10

SCIENCE AND POLICY

An understanding of complex environmental issues is likely to require contributions from many scientific disciplines. The best known and most ambitious attempt to bring different scientific disciplines together is the Intergovernmental Panel on Climate Change (IPCC), established in 1988 by the United Nations Environment Programme (UNEP) and the World Meteorological Office (WMO), and then given endorsement by the UN General Assembly. Its purpose was, and remains, to provide the world with a clear scientific view on the current state of knowledge on climate change and its potential environmental and socio-economic impacts. The creation of IPCC stimulated debate in academic circles about methods for such 'integrated environmental assessment' and how it could be extended to other subjects.

The European Forum for Integrated Environmental Assessment (EFIEA) was formed with EU funding in 1998 to contribute to this debate and to foster cooperation between scientists and decision-makers within the EU and beyond (Tol and Vellinga 1998). In the discussions that took place when EFIEA was being formed, and when reading the literature on the subject, I was struck by the way so many scientists could barely conceal their irritation that policy-makers appeared not to take account of what they were saying. The paper below, delivered at EFIEA's first workshop (Haigh 1998), was an attempt to shed some light on the confusing world of policy-making, and not just in the EU, for an audience with a deep knowledge of their own subjects but with rather limited knowledge of how policy is made. Its message was that if scientists are to contribute to the policy-making process, it is important to recognise that it operates in a quite different way from science. Scientists need some awareness of the rules (or lack of them) of the policy process just as much as policy-makers need to try to understand what scientists are telling them. I took two case studies to illustrate the unpredictability of EU policy-making. The first showed how lead was banned in petrol in the EU only because two different countries wanted, at much the same time, to solve quite different problems based on quite different scientific evidence. The fact that policy communities in different countries were speaking different languages, in more than the literal sense, would have made an 'integrated environmental assessment' very difficult to organise, and none was attempted. The second case study (emissions from small cars) showed how a decision was made not primarily as the result of scientific advice, but as a result of the convoluted procedures of the EU policy-making process. These exist to balance the interests of the EU Institutions and the Member States. The outcome was a surprise to many, but in that case the best environmental outcome was nevertheless achieved.

Science–policy interactions from a policy perspective – paper delivered in Amsterdam, March 1998

> You must remember that argument is constructed in one way and government in entirely another

I quote the words of Lord Macaulay, the nineteenth-century English historian and politician, in order to introduce at the outset a note at least of caution, if not of scepticism. I intend in this paper to give some examples to show just how unpredictable the process of European environmental policy-making can be – some people might think muddled is a better description. It seems necessary to emphasise this rather forcibly so that the ambitious enterprise of Integrated Environmental Assessment (IEA), and our European Forum that we have launched to promote it, do not fall into the trap of having exaggerated claims made for them, which in time are found to be wanting.

We certainly need to avoid giving the impression that some new tool has been invented that will somehow solve the problems of making policy. I welcome the modesty of some writers, who while conceding that the term IEA is new, acknowledge that the IEA is 'far from novel' and is 'an elaboration of existing techniques rather than a truly new area' (Wieringa 1996). While IEA is an attempt to provide policy-makers with fuller information on which to base their decisions, any claim to completeness, which is implied by the word 'integration', is always likely to be challenged by those whose interests will be affected by policy decisions. One can even go so far as to say that policy-making is the taking of decisions in the absence of adequate information and that when information is accepted as complete by all the affected parties then the decision does not require much skill.[1] IEA is certainly a welcome counter to the view that policy-making is largely about power and interests, a point made by Majone when he said that '… (we) miss a great deal if we try to understand policy-making solely in terms of power, influence and bargaining, to the exclusion of debate and argument' (Majone 1989).

The various definitions of IEA all insist not just on its comprehensive character but also on its policy relevance. Our Forum's 'living document' (Tol and Vellinga 1998) says it is to involve decision-makers as well as scientists. Whether decision-makers are merely the recipients of the product of IEA or are directly involved in its production – and this is a matter on which some clarification is still needed – the intention is certainly to influence policy rather than just to advance knowledge.

If IEA is to influence policy its chances of success will be greater if it is framed with political awareness and with some understanding of the policy-making process which has its own dynamics and logic, or lack of it, and which Macaulay reminds us is constructed in an entirely different way from the reasoned arguments of scientists and academics.[2] Once we are dealing with policy-making we cannot escape its unpredictable character and cannot assume that the rules which apply to some parts of IEA dealing with scientific knowledge can apply to the whole.

Even within the academic community, characterised as it is by such great respect for rational argument, we know that methods vary considerably. When the British biologist Eric Ashby spoke on reconciling man with the environment, he justified his temerity in giving lectures endowed to support the humanities in words that are very relevant to our subject:

> It seems to me that the basic difference between science and the humanities is not the choice of subject; it is the spirit of pursuit (a phrase the philosopher Samuel Alexander used in another context). The scientific spirit is one that tries to eliminate the human equation, so that the findings will be acceptable to any rational mind anywhere. The humanistic spirit of pursuit is one that does not try to eliminate the human equation; it tries to solve it, and every solution is unique. Every man converts the oxygen he breathes into carbon dioxide in the same way. No two men love or act justly or show disdain in the same way, either toward one another or toward nature. These lectures are about the influence of the sentiments upon man's attitude to the environment. I hope this qualifies them to be included in an endowment devoted to the humanities.
>
> *(Ashby 1978: v–vi)*

Ashby was here comparing two categories into which academic disciplines can be divided, but his lectures also dealt extensively with the processes of policy-making based on his own experiences. It is the differences and interaction between policy-making and science that I want to explore in this paper, rather than the difference between humanities and science. I doubt that the rough trade of policy-making would ever deserve to be called a branch of the humanities, although a study of the processes certainly can be. Nevertheless, I find Ashby's justification and lectures a helpful starting point although I will extend his analysis by adding another element that is missing for the very good reason that environmental policy-making in an international context was only in its infancy when his lectures were given. Now the international context cannot be ignored.

If no two men love, or act justly, or show disdain in the same way, neither do countries. A statesman or politician is driven by his own sentiments and will probably make no attempt to remove the human equation. If he is elected he doubtless strives to understand, and may claim to reflect, the sentiments of his electorate. A national government, composed as it is of human politicians and advised by human officials and responding to national pressures, is bound to reflect national preoccupations. This adds greatly to the interest and difficulties in the making of international environmental policy as well as in attempting to study it.

In approaching this subject in an international, or at least European, context I want to start with a frame provided by Ashby in the lectures I have mentioned. As well as being a biologist, Ashby was also a university administrator (Vice-Chancellor of Cambridge University), an adviser on policy (first Chairman of the Royal Commission on Environmental Pollution), and a parliamentarian (in the House of Lords) who took a great interest in EC environmental policy in its early years. Ashby drew on this experience in developing what he called the 'chain reaction'. One such experience related to waste policy. The Royal Commission had calmly said to the Government that the system for handling hazardous waste was badly in need of reform. The Government listened politely but had other things on its mind. Two years later the Commission repeated its concern and again the Government did nothing. Then two stories appeared almost simultaneously in the press, one about children playing on a rubbish tip with a drum marked 'cyanide', and another about a lorry driver being paid to dump waste illegally. This caught the public's attention and the Government responded by rushing the Deposit of Poisonous Wastes Act 1972 through Parliament in near record time.

This and other experiences of the failure of rational argument alone to achieve results led Ashby to formulate the chain reaction as follows:

> Between the disclosure of an environmental hazard and political action to control the hazard there is a complex chain reaction of events. The public conscience is excited; public-interest lobbies mobilise to defend the environment; self-interest lobbies set out to stop decisions that threaten their interests. Among those who have responsibility for dealing with the hazard there is an inbuilt inertia manifested as stalling questions: 'Is this just a passing scare?' 'Will the problem go away if we take no notice of it?' When it becomes clear that the problem will not go away, the responsible body (politician or administrative agency) seeks advice from experts in science and from experts in economics. In due course the experts give their opinions, the lobbies deliver their volleys of advocacy, and the politician has to make a decision that may ultimately be codified in law.
>
> This chain reaction is complex, but it can be simplified into three stages. In the first stage – let us call it the ignition stage – public opinion has to be raised to a temperature that stimulates political action. In the second stage the hazard has to be examined objectively, to find out how genuine and how dangerous it is, and just what is at risk. In the third stage this objective information has to be combined with the pressures of advocacy and with subjective judgments to produce a formula for a political decision. We can, therefore, think of the chain reaction as an initial ignition of the issue followed by inputs. The inputs

> of the scientist and the economist are assumed to be objective (some say 'value-free', but we shall come to that later), whereas the inputs of spokesmen for public and private interests are known to be subjective; to these is added the politician's input, which includes not only facts but 'hunch'. In the two lectures after this one I shall reflect on the value problems that confront the scientist, the economist, and the politician.
>
> *(Ashby 1978: 14–15)*

This three-stage reaction, rich though it is in insights, cannot be regarded as a complete account of the policy process since it does not deal with what happens after the political decision is made. Pieter Winsemius – perhaps without knowing of Ashby's three-stage chain reaction – has proposed what he called the Environmental Policy Life Cycle which adds a fourth stage, dealing with implementation, to three stages that are very similar to Ashby's. He called them: (1) recognition; (2) policy formulation; (3) solution; (4) administration (or 'maintaining control') (Winsemius 1986). Others have suggested that a sequence of overlapping policy cycles better describes the real world.

The reason I quote Ashby at such length is because he gives full weight both to the objective aspects ('the inputs of the scientist and economist are assumed to be objective') and to the obviously non-objective aspects ('the volleys of advocacy' and 'subjective judgements'). The first, or ignition, stage is manifestly not objective. Even in the second stage, the inputs of the scientist and the economist can only be 'assumed' to be objective. Embedded in the three-stage process is the recognition that the making of policy or 'government' is constructed in a different way from 'argument'. I find rather little recognition of this in the definitions of IEA quoted in EFIEA's 'living document'. While IEA is an attempt to bring together different academic disciplines in order to be more politically relevant than a single academic discipline can be on its own, the question must remain open about the extent to which IEA is to engage more fully in a process that involves so many non-academic aspects.

Ashby's three-stage chain reaction may not be complete, but even within its own terms is it correct? In particular, does there always have to be an ignition stage involving public opinion? I will attempt to answer this question by taking a case study of the issue of lead in petrol, an issue which first came to prominence in both the UK and Germany in 1971; involved two EC Directives; dealt with two very different scientific issues; and is still not finally resolved because we have yet to set a date when all petrol has no added lead. We can then speculate how differently the matter would have been handled if the tool of IEA had been well developed and applied in the 1970s and 1980s.

Case study – Different national perspectives

Figure 10.1 that elsewhere (Haigh 1986) I have called a "Majestic Descent" shows the permitted lead level in petrol in the UK from 1970 onwards. One could

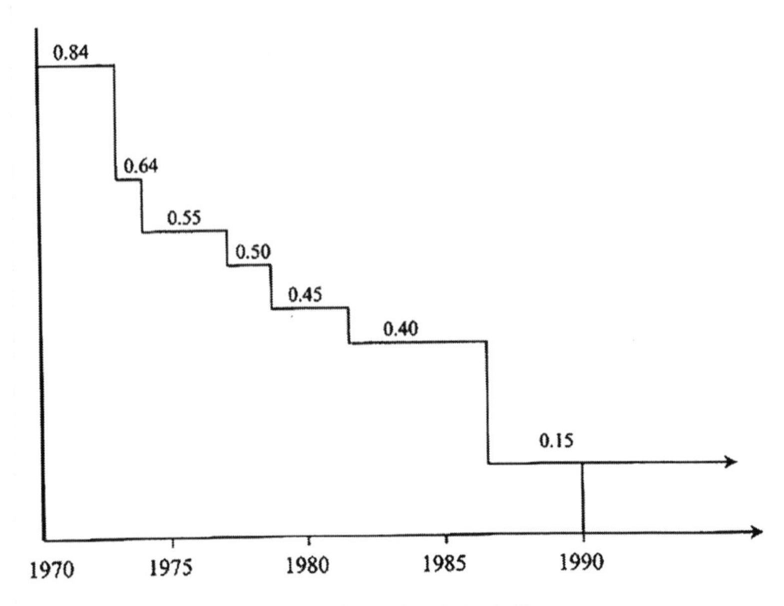

FIGURE 10.1 Permitted lead content of petrol in UK (g/l)

construct a graph for all other EU Member States, each slightly different but all showing a similar downward descent. How is one to explain this descent? Was it pre-planned by far sighted Ministers, advised by sober and well informed officials who had consulted all the relevant sources of scientific advice, biological, technical and economic? Notice that there are no precipitate leaps from 0.84 g/1 to zero. Notice how all is steady so that all affected interests can calmly adapt themselves. This is a caricature and the story is not like that. Or do we have here a picture of a terrible war, each downward step a pitched battle waged between the forces of enlightenment and the forces of reaction until in the right-hand corner the latter are finally crushed into the ground? It is not like that either. In fact my first caricature is not a bad description of the top left-hand corner of this picture nor my second of the bottom right-hand corner. There was certainly no grand plan based on complete scientific assessment.

Let us look at developments more closely in Figure 10.2. Here I have indicated a number of the key events that help explain the shape of the descent. I have put these events under five headings, some 'objective' and some not, one of them being a war outside Europe.

In the UK the story starts in 1971 when the Government received advice from its Chief Medical Officer that air lead levels should not be allowed to increase above levels then prevailing. The Government responded by deciding on a phased three-stage reduction from 0.84 to 0.4 g/l, to be achieved in 1976 in line with anticipated growth of car use. This phasing was then delayed and the deadline postponed until required by the EC Directive. While the plan was rational the execution was not as predicted.

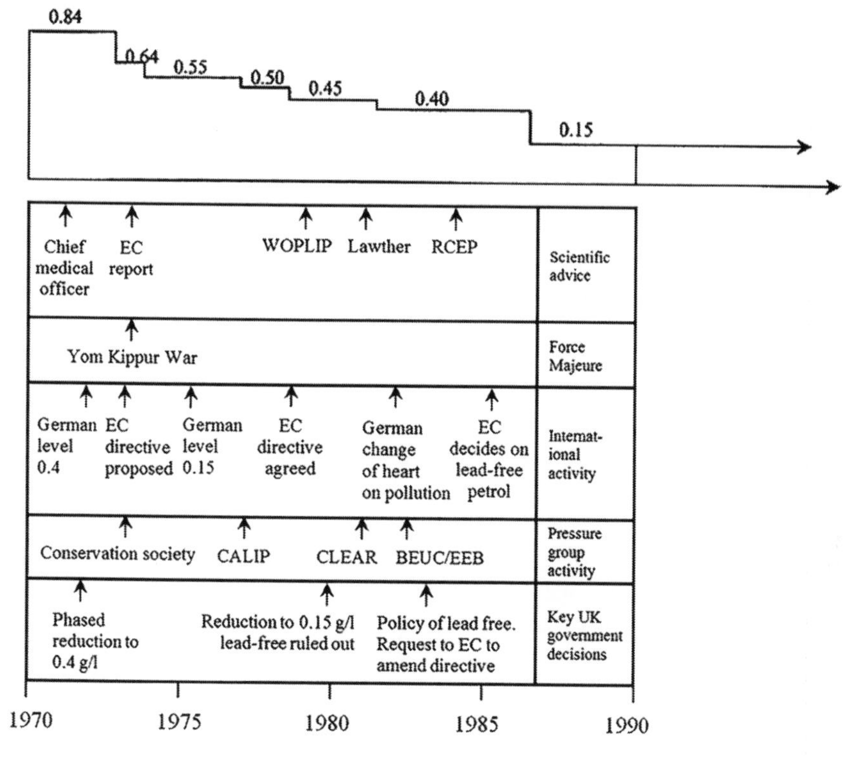

FIGURE 10.2 Key events which help to explain reduction in UK lead levels in petrol

Also in 1971 the German Government decided to reduce lead levels but chose a faster programme which was implemented as planned: 0.4 g/l in 1972 and 0.15 g/l in 1976. I have never looked at the scientific data that caused the Chief Medical Officer in the UK to issue his warning in 1971 and I do not know if it was the same that caused the German Government to act. But if it was – and it seems probable – note how two governments act differently. I am told there was no public pressure in Germany for these reductions which were decided purely on the basis of the scientific information. The earliest pressure group activity in Britain that I know of was the call for lead-free petrol by the Conservation Society in 1973 because of concern about effects on human health. (The Conservation Society later formalised this into the Campaign Against Lead In Petrol, CALIP, in 1977.)

So in both countries the ignition stage took place among scientific circles without much public excitation. One may ask why two governments took different approaches and one factor is likely to be that the largest European plant manufacturing the lead additives for petrol was to be found in the UK. When there are no, or weak, interests defending the status quo there may be no need for public pressure.

In 1971 the UK was not yet a member of the EC and it was the unilateral decision by Germany that resulted in the EC Commission establishing two committees in 1971 to study the health and technical aspects of lead pollution from motor vehicles. The work of these committees is summarised in the explanatory memorandum that accompanied the proposal for the first Directive issued in 1973 – COM(73)2050. The memorandum concluded that although there was no immediate danger for public health, it was desirable to prevent an increase of air pollution by lead and hence to limit lead because of the increase in car use. Other reasons were related to the common market since unilateral action risked making it more difficult for oil refiners to sell petrol across frontiers. Perhaps the work of these committees could be called IEA of a limited kind.

The Commission's proposal for a Directive was a modified version of the German plans. The first stage was to be a limit of 0.4 g/l, but the second stage of 0.15 g/l was not to apply until 1978 and was to be limited to 'regular grade' petrol, with 'premium grade' remaining at 0.4 g/l.

In the European Parliament (10 November 1975) the rapporteur of the Environment Committee said that the proposed second-stage reduction (to 0.15 g/l for regular grade) had 'met with insurmountable opposition in the Committee' because it would have involved the industry in substantial investment as well as increasing petrol consumption. 'Since', he went on, 'these objections could not be refuted, the Committee preferred to require the Commission to postpone the introduction of the second stage.' The limit of 0.4 g/l was, however, approved by the Committee. The Commissioner in defending the proposal referred to the studies that had been done and said:

> our proposals are built to the best of our or anyone else's ability on the probabilities presented by these studies – I say 'probabilities' because nothing beyond that exists, neither here nor a few hundred kilometres away in the Federal Republic of Germany.

He explained that the Commission was not convinced that it was necessary to go as far as the German Government was proposing, but pointed out that the delay in obtaining an opinion from the Parliament meant that it would not now be possible to obtain a Council decision on the proposal before the second-stage reduction took effect in Germany on 1 January 1976. All the subsequent discussion in the Council – where decisions then had to be taken unanimously – was therefore coloured by an existing German limit of 0.15 g/l. The Directive finally agreed in 1978 therefore had to allow Member States to introduce a national limit of 0.15 g/l but its main provision was an upper limit of 0.4 g/l.

Here we have a clear admission that the standard proposed was based on probabilities only, and with the result influenced by political exigencies. The figure of 0.15 g/l did indeed have a technical basis, but one unconnected with human health: it was close to the lowest level usable in some existing petrol-engined cars without special adaptation.

The reason the Directive was not adopted until 1978 (with the requirement for the new standards to be met only in 1981) was because of the Yom Kippur war and the resulting shortage of oil in Europe. It was believed at that time that more crude oil was required to produce a given volume of lower-lead than higher-lead petrol.

The story of this first Directive 78/611 shows how in favourable circumstances a determined Member State, in this case Germany, can – despite considerable opposition and scientific uncertainty – pull the rest of the Community along behind it so that higher environmental standards are achieved more quickly than if the Member States had proceeded at their own pace, even though it did not go as far as requiring lead-free petrol. The introduction of lead-free petrol can be regarded as a new and separate phase in our story. The second Directive 85/210 requiring this was proposed in 1984 and adopted much more quickly than the first. The original pressure for it came from the UK, but without a lucky coincidence it might never have been adopted.

In the UK in 1981, following the reports of a scientific committee, chaired by Professor Lawther, on lead and health (Lawther 1980), and of a governmental Working Party on Lead In Petrol (WOPLIP) (Department of Transport 1979), the Government decided two things: first, to require leaded petrol to be limited to 0.15 g/l, the lowest that could be required by the Directive and, second, to rule out lead-free petrol. This was after a major battle between Ministries: Health together with Environment were defeated on the second point by Transport, and Energy and the Treasury. The level of 0.15 g/l was the level adopted in Germany six years earlier. Then there was a dramatic change in policy. In April 1983 the Royal Commission on Environmental Pollution recommended (RCEP 1983) that the Government should initiate negotiations with the Commission and other Member States to secure removal of the lower limit of lead in petrol in Directive 78/611 so that at the earliest practicable date all new cars should be required to run on lead-free petrol. The Government immediately accepted this recommendation.

Between these two decisions in the UK (1981 and 1983) there was an extraordinary public campaign. A new organisation called CLEAR (Campaign for LEad-free AiR) supported by a millionaire with a mission provided very effective political lobbying and also publicity for the scientific information. It is possible that the Royal Commission only decided to look at the issue of lead because of the campaign, although that is not the view of its Chairman (Richard Southwood). What can be said with some certainty is that the Government only endorsed the Commission's conclusions so quickly (within half an hour of publication) because of the campaign and because of an imminent general election. So here we have a perfect example of Ashby's ignition stage 1, but in this case superimposed on the objective evaluation stage 2 (the reports of the Lawther Committee and the Royal Commission). But before action could follow the EC Directive would have to be amended.

When the first approaches were made to the Commission in April 1983 the reaction was negative. Initially the attitude of Commission officials – one shared in Germany which felt that the problem had been solved some years ago – was that Directive 78/611 was sufficient and that the other Member States were unlikely to be persuaded by the new, and uncertain, evidence about damage to children's brains. The coincidence that changed the debate was the new found concern in Germany about forest die-back. Germany realised that to achieve its objective of significant NO_x

[nitrogen oxides] reductions from cars, catalytic converters would be required, and that as these are poisoned by lead, then lead would have to be removed. It has always surprised me that Germany, which in June 1982 had submitted a memorandum to the Council about reducing emissions from industrial plant, did not at the same time press for lead-free petrol to reduce NO_x. That only came a year later after the UK was arguing the case for public health. Germany, together with the Netherlands and Denmark, then supported the UK initiative, but for quite different reasons. After French resistance was overcome the new Directive 85/210 was then adopted.

So here we have an issue with an ignition stage in both UK and Germany, but with public conscience being excited about two quite different issues: public health, and death of forests. It is pure chance that they came together at the same time and if either had been missing it is quite possible that the Directive would not have been agreed, or not agreed so quickly. And, if we are tempted to speculate further, what would have happened if someone had invented a lead-tolerant catalytic converter?

We can say with certainty that there was no integrated environmental assessment that brought the two issues together at an early stage. They were only brought together after two governments had already decided their policies and found themselves both pressing the Commission to propose a Directive, but for different reasons.

In order for there to have been an IEA, the problem would have had to be defined at the outset as 'reducing damaging vehicle emissions' instead of 'health effects of leaded petrol' and 'effects of vehicle emissions on forests'. It would have been necessary to have brought together very different scientific disciplines and would have involved other considerations such as the effects of the additives that replaced lead. Probably the greatest obstacle to establishing an early IEA would have been the difficulty of linking lively debates that were taking place in two different countries on two different topics. Two different policy communities urged on by different publics excited by different issues, speaking different languages, and seeking to influence different governments would not easily have wanted to redefine the problem. It might have been easier to bring them together at a European level, rather than just in Germany or the UK, but the Commission was not then (in the early 1980s) mature enough to have made the connection at an early stage, nor did any other European body do so.

Case study – Procedure influencing outcome

I want now to give another example of environmental policy-making in the EU in which the choice between possible outcomes, each involving different technologies and with different environmental effects, was influenced not so much by scientific considerations as by the procedures of the decision-making process. It falls firmly within Ashby's stage 3 – the production of a formula for political decision. It follows on from our first case study since it concerns the making of the decision in 1988/89 that from 1992 all small cars should have catalytic converters, a decision which would not have been possible without the availability of unleaded petrol.

The issue was a choice between three technologies, each involving different costs and different benefits. One technology – the 'lean-burn' engine – while emitting more NO_x consumed less fuel, that is, it emitted less CO_2 and the decision therefore involved a trade-off. The other technologies were the oxidation catalyst and the more expensive three-way catalyst.

The limit on combined emissions of HC [hydrocarbons] and NO_x of 19 g per standard test had already been set for 1988 and a more stringent standard of 15 g per test had been agreed for 1990. What now had to be decided was the standard for 1992, and there were effectively three possibilities (12, 8 or 5 g per test) depending on the technology chosen. These are summarised in Table 10.1.

The process of reaching the decision in the EC involving the Commission, the Parliament and the Council was complicated by the 'cooperation procedure' newly introduced by the Single European Act.

The procedure, summarised in Table 10.2, affected the small cars decision as follows.

TABLE 10.1 Emissions from small cars – decision-making in 1988/89

Year	*Status*	*Standard*
1988	Standard in force	19 g per test (HC + NO_x combined)
1990	Standard already agreed	15 g per test
1992	Possibilities	12, 8 or 5 g per test

5 g per test could only be met by using a three-way catalyst with electronic management. This cannot be used with a 'lean-burn' engine and is more expensive than an oxidation catalyst.

TABLE 10.2 Cooperation procedure (summary of Article 189c of the Treaty of Rome)*

Step	*Action*
1	Commission proposes legislation
2	Parliament adopts opinion by simple majority ('first reading')
3	Council adopts 'common position' by qualified majority voting (QMV)
4	Parliament approves 'common position' or amends/rejects it by absolute majority of those entitled to vote ('second reading')
5	Commission may amend proposal to take account of Parliament's amendments
6	When Parliament has proposed amendment the Council may: • adopt by QMV amended Commission proposal; or • adopt unanimously Parliament's amendments not approved by the Commission; or • amend and adopt unanimously Commission's proposal.
7	If Parliament rejects proposal, the Council can still adopt it unanimously.

*This procedure (cooperation between the Council and the Parliament) was introduced by the Single European Act in 1987. The Treaty of Maastricht has extended it with the 'codecision procedure' which further increases the Parliament's power. The Treaty of Amsterdam will make the 'codecision procedure' the norm for all environmental legislation.

1. The Commission started the procedure by proposing 8 g per test. Some governments (including the UK) wanted 12 g as this would allow development of the lean-burn engine, but others (including Germany and the Netherlands) wanted 5 g. Opposition to 5 g came from the small-car producing countries (France and Italy) on grounds of costs. They feared the high cost of three-way catalysts would reduce the sale of their small cars.
2. At its first reading in September 1988 the Parliament wanted 5 g. The vote was 243 for, 63 against, and 14 abstentions. (This was a substantial majority but not an absolute majority of those entitled to vote since there are over 600 MEPs, but at 'first reading' a simple majority is sufficient.)
3. Early in 1988 it became apparent in the Council that 12 g would not find a qualified majority. The supporters of 12 g then changed their position and compromised by supporting the Commission's proposal of 8 g. A 'common position' in favour of 8 g was then formally adopted in the Council on 24 November 1988 having been agreed informally in June during the German Presidency. This was achieved when the German Presidency (Klaus Töpfer) moved reluctantly but in a statesmanlike manner (as befits the President) from supporting 5 g to supporting 8 g in order to break a stalemate.
4. At its 'second reading' on 11 April 1989, the Parliament reaffirmed its 'first reading' position and by an absolute majority amended the Commission's proposal to 5 g (311 for, 5 against, and 5 abstentions).
5. Before this vote was taken the Commission let it be known that it would amend its proposal to 5 g if the Parliament were to propose such an amendment. This was a change of policy following the appointment of a new Commissioner for the Environment – Ripa di Meana – who had decided to take a progressive position in environmental policy, and could see popular support.
6. In May 1989 the Commission formally amended its proposal to 5 g, taking account of the Parliament's amendment.
7. The Council on 9 June 1989 could not agree by unanimity to change the proposal back from 5 to 8 g. There were therefore two possibilities, either to adopt 5 g or to agree nothing, leaving the standard at 15 g. The Council agreed 5 g.

One moral from this story, and why it is told here, is that the outcome was determined by the procedure. If the decision had been taken before the Single European Act introduced qualified majority voting (QMV) in Council it is possible that 12 g would have been the outcome. Any one Member State could have vetoed the others, and if one Member State had insisted on 12 g the others might have felt 12 g preferable to leaving the standard at 15 g. If, on the other hand, there had been QMV but no cooperation procedure with the Parliament, then the outcome would have been 8 g. With QMV and the cooperation procedure the result was 5 g. So different procedures would have produced different results quite independently of the technical possibilities, the economic consequences and the environmental effects. These of course were by no means ignored but had been balanced in different ways by different countries in adopting their negotiating positions. In the end it was the procedure that determined the way the different positions of the Member States were reconciled.

Application of IEA to EU policy-making

The EU affects the environment in two distinct ways: by adopting legislation that is primarily environmental (e.g. standards for emissions to air or water); and by adopting other policies that have an impact on the environment (e.g. transport, agriculture). Since the Treaty of Rome states [Article 130r(2)] that 'Environmental protection requirements must be integrated into the definition and implementation of other Community policies', EU policy mirrors that in most countries, in that it is effectively composed not only of what is traditionally the field of Ministries of the environment, but also of policies of other Ministries, possibly under the stimulus of the environment Ministries, to advance the cause of environmental protection. These two elements inevitably overlap and both provide scope for IEA.

Possible EU environmental policy can be discussed and proposed in general terms by the European Commission in documents such as an Action Programme, but to become binding on the Member States such policies must be embodied in items of legislation adopted by the Council and Parliament (Directives, Regulations or Decisions). Over the years several hundred items of environmental legislation have been adopted, a couple of which we have described above, and these can be grouped under the following headings:

- water
- air
- waste
- substances
- noise
- radioactivity
- wildlife
- climate change.

A recent Directive seeks to integrate pollution of the different media caused by industrial plant and in addition there are cross-cutting Directives to underpin environmental policy such as the Directive on environmental impact assessment of development projects and a Directive on freedom of access to environmental information.

The Fifth Action Programme on the Environment of 1993, which runs till 2000, in considering other policy areas that have major impacts on the environment, selected the following five target sectors:

- industry
- transport
- agriculture
- energy
- tourism.

If IEA is to develop at a European level it will eventually have to contribute both to the field of traditional environmental legislation and to the field of the other policies that affect the environment. IEA is already contributing to the proposed acidification strategy which falls under the heading of air pollution control, and also to climate change which falls both under the heading of 'pure' environmental policy and also of energy policy. The target sectors selected for the Fifth Action Programme may be added to as the next Action Programme comes to be prepared, for example, by including fisheries. Policies for these topics will develop whether or not IEA is formally applied, and our Forum will have to decide to which it can best contribute.

Assessment

The two case studies above show that environmental policy-making extends beyond the scientific identification of an environmental problem and the rational selection of the best environmental solution, however sophisticated or simple that part of the process may be. The process is complicated by the need for ignition of the public conscience so that the temperature is raised sufficiently for politicians to feel able to take those decisions that may hurt some economic interests. Science plays a unique and essential role in informing the public and influencing and guiding public opinion, which is a major determinant of policy. But science itself does not always reach the public at a specific point in time when a specific decision is called for. Since we cannot yet claim that there is a European public, but only a collection of national and regional publics, the way the policy debate develops under the pressure of public opinion, more or less informed by science, is very likely to differ between countries. European policy-making is very much about reconciling these differences, and to the extent that IEA can contribute it will be welcome.

The procedure for making decisions also affects the decisions, and procedures vary from country to country. The example given above concerns EU procedures but each country has its own traditions, embodied in its constitution and governed by differing laws. Surprising as it may seem to scientists these differences do affect the content of policies. Although rational scientists may speak a universal language ('every man converts the oxygen he breathes into carbon dioxide in the same way'), practitioners of administration and law in different countries do not speak the same language and nor does the public. Environmental policy-making in a European context brings different administrative and legal cultures into contact with each other, sometimes with some friction, and often with the participants not really knowing the full reasons for each other's positions. The first step is to acknowledge, preferably without irritation, that such differences exist. The next step is to understand them, and we might decide that for European policy-making this is a necessary part of IEA and gives EFIEA a special role. We may on the other hand decide that IEA cannot really cope with this, but any attempt to play down or even eliminate the human equation will doom IEA to irrelevance.

IEA is certainly an attempt to refine and strengthen Ashby's second stage – that of objectively examining the problem. It is not clear to what extent it can also provide a contribution both at stage 1 (the ignition stage), either by helping to identify the problem or by helping to excite the public conscience, and also at stage 3 (the combination of objective information with the pressures of advocacy and with subjective judgement to produce a formula for political decision). Stage 1 may feel a little too like popularisation for the more fastidious scientist and stage 3 may be much too much like politics, which indeed it is. We must either agree to define the boundaries of IEA more clearly, or we must be prepared to accept uncertain boundaries. If we agree that IEA is confined to stage 2 we will certainly have simplified matters but the result will be a rather incomplete contribution to the making of policy. If we try to include stages 1 and 3 we are in for trouble from those who think that the only things that matter are those that can be measured objectively.

The above discussion of lead in petrol showed European institutions not then really able to adopt an IEA approach. In the late 1990s we have moved on and it is possible that the European Environment Agency can now play this role, and perhaps EFIEA will come to make more of a contribution. I do not say it is impossible to undertake an IEA in all circumstances but we should not delude ourselves that it will always be easy. When either the scientific community in several countries, or the public, is seeing a problem from the same viewpoint a solution is easier. The example of lead in petrol is an extreme case of a problem being perceived quite differently in different countries. Fortunately in that case the same solution was called for. One of the reasons why negotiations in Council are so interesting is precisely because the same solution, or the same pace, is not always called for.

After reading some literature on IEA I am left with the feeling that underlying it there is perhaps the thought that if only environmental problems were left to the rational scientists, and all the subjectivity associated with policy-making was eliminated, then better decisions would be made. But if the above analysis is correct, and if public excitation to raise the temperature sufficiently to take decisions is essential, then so is the political process, not just because politicians ultimately take the decisions but because in democracies politicians are our surrogates for public opinion. Politicians will not thank these proponents of IEA who offer them solutions which either they cannot deliver because they are unacceptable to public opinion or which have eliminated or played down public concern. This means that IEA must be framed with as much understanding as its proponents can manage of all the many complexities of policy-making. They must also be prepared to admit that any IEA is likely to be incomplete. The effort to be comprehensive is laudable; the claim that IEA will provide a complete solution would be misleading.

Paper presented and published in 1998

Notes

1 For some reason I did not refer to the role of the 'precautionary principle' in the taking of decisions under scientific uncertainty. For a discussion see Chapter 13.

2 There is a difficulty about the use of the word 'scientist'. In Anglo-Saxon usage the word is usually taken to refer to 'natural scientists' (biology, chemistry, physics, etc.) unless qualified. Economists do not normally call themselves scientists unless the context requires it, but do regard their discipline as 'scientific'. In mainland Europe the word 'scientist' is often used more broadly to mean 'academics'.

References

Ashby, E (1978) *Reconciling Man with the Environment*, London: Oxford University Press.

Department of Transport (1979) *Lead In Petrol: An Assessment of the Feasibility and Costs of Further Actions to Limit Lead Emissions from Vehicles*, London: HMSO.

Haigh, N (1986) Public perceptions and international influences. In: Conway, G, ed., *The Assessment of Environmental Problems*, London: Imperial College.

Haigh, N (1998) Challenges and opportunities for IEA – science–policy interactions from a policy perspective, *Environmental Modeling and Assessment*, 3(3), pp 135–142.

Lawther, P (1980) *Lead and Health: The Report of a DHSS Working Party on Lead in the Environment*, London: HMSO.

Majone, G (1989) *Evidence, Argument and Persuasion in the Policy Process*, New Haven, CT: Yale University Press.

RCEP (1983) *Lead in the Environment*, 9th Report, London: HMSO.

Tol, R and Vellinga, P (1998) The European Forum on Integrated Environmental Assessment, *Environmental Modeling and Assessment*, 3(3), pp 181–191.

Wieringa, K (1996) Towards integrated environmental assessment supporting the Community's environmental action programme process, ESEE Inaugural International Conference 'Ecology, Society, Economy', May 1996.

Winsemius, P (1986) Co-ordinating standards to avoid cross-media effects. In: *Proceedings of a Conference on Protecting the Public Against Hazards from Chemicals*, Brussels, Institute for European Environmental Policy/Bureau of National Affairs. Bonn: IEEP.

11

VOLUME CONTROL FOR SUSTAINABILITY

The phrase 'volume control' was used by the Dutch writer Wouter van Dieren at a conference I attended in Germany when he argued that environmental policy needed to move away from reliance on what he called 'quality control' and instead to 'volume control'. That was many years ago, but the phrase lodged in my memory, and some years later I used it in the title of a talk I gave to lawyers in London.[1] The talk, rather portentously called 'Space, time and volume control – further reflections on the law for sustainable development', dealt with the kind of legislation that was needed in helping to move towards sustainability. The talk largely discussed British legislation that used the words 'sustainable development' so is not suitable for a book about the EU. I repeated some of it at a later conference (Haigh 2003) and this chapter draws on both. It also draws on the deliberations of the NSCA's[2] Commission on Industrial Regulation and Sustainable Development, on which I served (NSCA 2001).

When John Gummer was the UK Secretary of State for the Environment he said we really must find better words than 'sustainable development' if people are to understand the concept. His own offering was 'not cheating on our children'. Another formulation is the one offered by a journalist who said, when someone attempted to explain it: 'Oh, I see, it means treating the Earth as if we are here to stay'. Both have advantages over the conventional explanation about achieving a balance between social, economic and environmental considerations that one hears so often. Both convey the tough element of the Brundtland definition (Brundtland Commission 1987) that pursuit of welfare today must respect the needs of future generations (see Chapters 1 and 3). Both say it relates to *time*, but the journalist's version adds *space*: our concern is not just with our own locality but must extend over the whole Earth. Indeed one cannot conceive of sustainable development being achievable at all below the level of the global system. No locality, or industry, or country can be sustainable on its own, since all are dependent on other parts of the Earth for resources or markets, and the most that each can do individually is 'to contribute towards the objective of achieving sustainable development'.[3] One question that was under discussion when I spoke in 2001 was whether sustainable development, when embodied in legislation, could by itself influence practical decisions such as whether or not to authorise a fossil-fuelled power station. Could it be relied upon in a Court to challenge such a decision, or does it just indicate a direction of travel? Is it just an aspiration, however world-changing that aspiration might be?

We know too well that some environmental problems are global and so require concerted action but many others are more circumscribed in space. Some issues, such as noise, or whether a waste site is well managed, are purely local, and indeed most environmental problems were regarded as local and within living memory most were handled locally. But certainly since the great UN Conference at Stockholm in 1972 it has been realised that many problems are regional and some are global. In theory it should be possible to identify the geographical extent of an environmental problem and we try to do so in order to tackle it. Examples where this has been done are given for air pollution in Chapter 4 and water pollution in Chapter 5. Sustainability moves our attention away from purely local problems to ones that extend in space. One consequence is that certain localities or countries may be required to take action for the benefit of others. Spain, for example, found itself agreeing to emit less SO_2 under the 'large combustion plants' Directive in 1988 (see Chapter 4) largely as part of an effort to reduce acidification of Swedish lakes and German forests, even though it hardly experiences acidification itself for reasons of geography, wind direction and soil conditions.

Similarly, sustainability moves our attention away from short-term problems to long-term ones. We have to think about future generations, and hence about time. Being prepared to make sacrifices today for someone else's benefit must be regarded as one of the characteristics of sustainable development, with the beneficiaries being displaced from us in *space* or *time* or both. To reflect this, the phrase 'inter-generational equity' has come into use, and 'solidarity between generations' is even to be found in the EU Treaties – see Chapter 3.

Whereas the geographical extent of an environmental problem is capable of being defined, its extent in time is indeterminate and we try to avoid thinking about it. Do we really mean sustainable in eternity? Probably not, since astronomers tell us that one day the Earth will fall into the Sun. If not forever, then for how many generations? We think about our children and grandchildren, but it is harder to think many more generations ahead. To make sacrifices for one generation may be quite different from making them for many. Because we cannot answer the question about how far ahead we should look – and no court of law would relish having to do so – sustainable development is best regarded as an aspiration rather than a rule. If one purpose of good law is to provide certainty, then sustainable development as a legal term fails the test resoundingly. If that is so, we need to find a new kind of legislation that is different from what was developed to deal with largely local and acute problems if we are to make a reality of sustainable development.

During the deliberations of the NSCA Commission mentioned above, an experienced industrialist gave an illuminating perspective on the evolution of environmental policy. He saw three phases under the catchwords 'safety', 'management' and 'sustainable development'. During the first phase, the only concerns of the industrialist were to avoid obvious damage. In the next phase, good industrialists sought to improve their environmental performance and minimise their effects overall, even in the absence of obvious damage. They were prepared to take action voluntarily. Now we were in the third phase, when industrialists are enjoined to pursue sustainable development and there is some confusion as to what extra is expected of them. One could only sympathise because thinking of the needs of future generations may well pull in a different direction from the immediate self-interest of most industries. The good industrialist will act voluntarily against his economic self-interest only so far. A new policy frame, supported by the appropriate legislation, is required which does not pull the good industrialist apart.

In Chapter 1 we argued that 1987 was an appropriate year to mark the beginning of the shift in EU environmental policy from obscurity to centrality. Among the reasons given for the choice of 1987 was the Brundtland Report of that year, which gave currency to 'sustainable development'. Once the concept was generally accepted, all policy-making and the legislation that gave it effect had to evolve to take account of long-term and long-range problems. National environmental policies which previously dealt with local and acute issues would need to be supplemented. The EU, by its very nature, can be expected to play a role since it has two great advantages. It is a grouping of many countries that together extend over a large geographical area, and it has the ability to think further ahead than most national governments since it is less constrained by the electoral cycle that forces national Ministers to focus on delivering results before the next election. Since national governments, under the principle of subsidiarity, do not like the EU to interfere in local issues, it has another reason to act strategically and so develop long-range and long-term policies.

The tool for environmental policy that van Dieren called 'volume control' has been developed – though the term is not generally used – for dealing with issues that are not primarily local and where the consequences are long term. It is therefore no surprise that it has been developed in the EU, though the EU is not alone in using the concept.

An early, and the clearest, example of its use is the 1980 EU Council Decision 80/372 that placed a cap on the production capacity for chlorofluorocarbons (CFCs) to counter the destruction of the ozone layer (see Chapters 1, 2 and 13). The volume control knob was in that case later turned down to zero. That happened when Regulation 3322/88 – and subsequent amendments – banned the production of CFCs altogether in order to implement the 1987 Montreal Protocol. The restricting or banning of the sale of harmful substances or products had long been known, but this was the first time that *production* was restricted. The response of the USA and of the 'Toronto group' of countries that were involved in the international negotiations was different. Their first response was to propose that well-established tool of environmental policy, namely a ban on the sale of certain products – in this case aerosol cans containing CFCs. The EU stuck to its guns by pointing out that this did nothing to control other uses that were growing, and insisted that its idea of a cap on production was the right way forward. It won the argument and so volume control became embedded in the Montreal Protocol. The originality of this step, and the EU's role in insisting on it, deserve to be better recognised.

The idea of controlling the total volume of emissions nationally, as opposed to production, of a harmful substance had come a little earlier. In Chapter 4 we saw how, in 1977, when the UN Economic Commission for Europe (UNECE) began developing the Convention on long-range transboundary air pollution, the Scandinavian countries had proposed a 'standstill' clause and a 'roll-back' clause for emissions of SO_2. This was formalised in the 1985 Helsinki Protocol which required parties to cut their national emissions by 30 per cent by 1993 compared with 1980. All parties accepted the same reduction, but this concept was refined when in 1988 the EU adopted the 'large combustion plants' Directive 88/609 which allocated different reductions to different countries depending on their circumstances. The concept of volume control then became the cornerstone of the climate change Convention, first with the aspiration to cap CO_2 emissions at 1990 levels by 2000 which the EU had proposed. The Kyoto Protocol then went on to set different cuts for different countries following the precedent of the 'large combustion plants' Directive (see Chapter 9).

The concept of volume control is essential for the operation of any emissions trading scheme, as the term 'cap-and-trade' implies. The EU emissions trading system, described in Chapter 9, depends on a cap on total national emissions of greenhouse gases. Volume control probably first developed in the USA where emissions trading in SO_2 became established. An imaginary 'bubble' was drawn around a given area with a cap placed on the total amount of pollution from any source allowed to enter into that bubble. Emitters could then trade among themselves so that the most economically efficient reductions should be achieved. In the US case the 'bubble' covered only limited areas within its territory, whereas with the Helsinki Protocol, the 'large combustion plants' Directive, and the Kyoto Protocol the areas covered are the whole of the relevant nation states. Volume control has thus become a major tool for dealing with international environmental issues. As noted in Chapter 4, the word 'national bubble' was used by analogy with the US example as this new tool evolved. Volume control is perhaps a more weighty phrase than a 'bubble' to express this important concept.

There are other examples at EU level of the use of volume control. The 'national emissions ceiling' Directive 2001/81 sets out to do exactly what its name implies. Different ceilings have been set for the different Member States for several air pollutants. The 'water framework' Directive 2000/60 limits abstraction of water to the rate of recharging. The 'landfill' Directive 1999/31 limits the amount of biodegradable waste that may go to landfill. Nor is the concept confined to traditional environmental policy. The fishing quotas established under the Common Fisheries Policy and 'set aside' under the Common Agricultural Policy – under which a certain percentage of agricultural land was withdrawn from agriculture – can also be regarded as a form of volume control.

At national level one of the most ambitious examples of volume control is to be found in the UK Climate Change Act 2008. This sets a national cap on net emissions of greenhouse gases at 80 per cent below their 1990 levels by 2050.

From these examples of 'volume control' being applied in practice, we can define it as *legislation (or an international convention) influencing the total volume of some activity as opposed to controls applied only at site level.* Of course the volume control may need to be translated into controls at site level, in which case it provides a strategic level of control over more local decisions. Volume control, then, is a key way in which the concept of sustainable development can be translated into laws that bear on long-range and long-term environmental problems.

It is always dangerous to speculate ahead as to what developments are to be expected, but there is already discussion about leaving known reserves of fossil fuel untouched, and about limiting air traffic by not expanding airports. The 7th Environmental Action Programme has talked of 'targets for reducing the overall environmental impact of consumption' and greater moves towards achieving a 'circular economy'. All these could fall under the heading of volume control.

When in 1987 the Montreal Protocol was adopted, and when in 1988 the EU adopted the 'large combustion plants' Directive, it was widely recognised that both were something new, so there was nothing special about the article reprinted in Chapter 4 being called 'New tools for European air pollution control'. Since then the concept of 'volume control' has evolved, most notably for climate change. It now needs to be recognised that the EU has devised a new tool that contributes to the achievement of sustainable development.

Notes

1 At the Annual General Meeting in 2001 of the UK Environmental Law Association (UKELA).
2 The National Society for Clean Air (NSCA) changed its name to Environmental Protection UK (EPUK).
3 This convoluted wording comes from Section 4 of the UK Environment Act 1995 which sets out the aim of the Environment Agency (England and Wales). In my 2001 paper at the UKELA meeting I discussed what that Section entailed. I had just ceased to be a Board Member of the Environment Agency.

References

Brundtland Commission (1987) *Our Common Future: Report of the World Commission on Environment and Development*, Oxford: Oxford University Press.

Haigh, N (2003) Sustainable development in the European Union Treaties and in national legislation: some conclusions. In: Fitzmaurice, M and Szuniewicz, M, eds, *Exploitation of Natural Resources in the 21st Century*, Alphen aan den Rijn, the Netherlands: Kluwer Law International.

NSCA (2001) *Smarter Regulation: The Report of the NSCA Commission on Industrial Regulation and Sustainable Development*, Brighton: National Society for Clean Air.

12

ALLOCATING TASKS – SUBSIDIARITY

The allocation of tasks between different levels of government is well understood in all countries and is sometimes a source of conflict. The allocation of tasks between the EU and its Member States has also given rise to conflict, and one of the innovations of the 1992 Treaty of Maastricht was the introduction of the principle of 'subsidiarity' intended to ensure that the EU took action 'only if … the objectives of the proposed action cannot be sufficiently achieved by the Member States'.

The EU can act only if the Treaties give it the power to do so, that is, if it has the 'competence' to do so. In the EU, the principle of subsidiarity is applied to determine not whether the EU has the competence to act, but whether the matter in question might better be handled by the Member States even though the EU has competence.

The Treaty of Maastricht was controversial in many Member States for many reasons. It is probably best known for three innovations: for creating the 'European Union' (with the EC for a time as one of its pillars); for preparing the way for the single currency – the euro; and for increasing the powers of the Parliament. At that time there was criticism of the EU in several countries for interfering too much in national affairs, and the newly introduced principle of 'subsidiarity' was intended to allay this and accordingly generated much discussion. A conference in Brussels in 1993 considered its possible effects on environmental policy – it was organised by the European Society for Environment and Development (Dubrulle 1994). I presented the paper reprinted below and repeated it at a conference at the London School of Economics (Haigh 1994).

The environment as a test case for subsidiarity – paper delivered Brussels, October 1993 and London, February 1994

The narrow vote in the Danish referendum on the Maastricht Treaty unleashed a debate on the deficiencies of the EC including its tendency to centralisation and interference in local affairs. The French referendum, the long drawn-out discussions in the British Parliament, and the German case before the Constitutional Court have all reinforced this soul searching about the future direction of the EC. One result was to focus discussion on the concept of subsidiarity that appears in Article 3b of the Treaty.

For many policy areas the concept of subsidiarity was new, but for EC environmental policy subsidiarity has been a formal element since its beginnings in 1973. The very nature of environmental policy ensures that this must be so.

Rubbish in the streets is a matter of environmental policy. So is the building of a factory or of roads. So are the condition of the Mediterranean and the subject of acid rain. So is the thinning of the ozone layer. Some of these subjects can only be handled by collaboration between countries but some can be dealt with perfectly well by national governments alone and some can best be dealt with by local authorities. It is obvious that some principle must guide the allocation of tasks, but a moment's thought reveals the complications.

Yes, the ozone layer is a global matter, and, yes, there is an international Convention on the subject. But practical action has to be taken at the level of the individual producer of CFCs [chlorofluorocarbons]; at the level of national authorities which must supervise exporters; at the level of local authorities which must collect old refrigerators; and at the level of the individual consumer. The EC had to be involved because different national approaches would distort competition. Effective action requires 'shared responsibility' between different levels, a concept usefully developed in the EC's Fifth Action Programme on the environment.

The question of the appropriate level for action has for long been an issue within countries. As an example let me take my own country, Great Britain, which can claim industrial pollution among its inventions. In the 1860s and 1870s there was a lively debate as to whether the control of air pollution should be exercised at national level or at local level. There were two schools of thought. One school believed in a national inspectorate on the model of the factory inspectorate established under the Factories Act 1833. The other school believed that local authorities should take responsibility, as they did for most other public health matters under the Public Health Act 1845. A typical compromise resulted: local authorities were to be responsible except for industries that were technically complicated. For these a highly qualified inspectorate was required and this was best done at national level. Shared responsibility (between a national inspectorate and local authorities) remains the basis for air pollution control in the UK today but of course with the benefit of EC legislation as well. Let me be clear that EC legislation has been a benefit both by introducing important new concepts, and by setting deadlines for action.

During that debate in the 1860s a Royal Commission produced the following description of subsidiarity (though the word came later):

> Local administration under central superintendence is the distinguishing feature of our government. The theory is that all that can be done by local authorities should be done by them, and that public expenditure should be chiefly controlled by those who contribute to it. Whatever concerns the whole nation

> should be deal with nationally, while whatever concerns the district must be dealt with by the district.

This is a useful statement precisely because it does not hide the difficulties. All that can be done by local authorities should be done by them, but they are not necessarily to be left on their own because central government can support them by 'superintendence'. Where public expenditure is involved, particularly if raised nationally by taxation, then central government certainly has a duty to control the expenditure.

Some of these points are relevant in an EC context. When Spanish roads are built with EC money, the EC has a right to know how the money is spent. The idea that the higher level has a responsibility to support the lower levels is particularly important. The word 'subsidiarity' derives, after all, from the Latin *subsidium* which means 'support' or 'body in reserve'. The higher level should not interfere with what can be dealt with perfectly well by the lower level, but it can help. In practice the question is whether the helping hand is a light hand or a heavy hand ('proportionality' or 'intensity' is the jargon). How light or heavy is often seen as the difference between support and interference.

One way that the national level can help the local level is by legislation, and similarly the EC can help Member States by legislating when the matter is sufficiently important and when for whatever reason the Member States do not do it themselves.

Similar ideas of course existed in all countries so when the EC adopted its First Action Programme on the Environment in 1973, the concept of subsidiarity was stated as one of eleven principles in these words:

> In each category of pollution, it is necessary to establish the level of action (local, regional, national, Community, international) best suited to the type of pollution and to the geographical zone to be protected. Actions likely to be most effective at Community level should be concentrated at that level; priorities should be determined with special care.

When the Single European Act was drafted in the mid-1980s and a Title on the Environment was introduced into the Treaty of Rome, the principle was restated in rather different words:

> The Community shall take action relating to the environment to the extent to which the objectives referred to in paragraph 1 can be attained better at Community level than at the level of the individual Member States.
>
> *(Article 130R: 4)*

These words later provided the model for Article 3b of the Maastricht Treaty.

So when the debate on subsidiarity began in earnest in 1992, those of us who had been involved in EC environmental policy were able to say – and we did – that EC policy on the environment has been guided by the principle all along.

One must observe that EC environmental legislation has not been confined to the two classic categories which are usually used to justify action at international level, i.e. transboundary effects, and distortion to competition. Drinking water and bathing water for example hardly fall into these categories, yet I would argue that the EC has on balance – and not without some friction to be sure – helped Member States to advance their policy on these important subjects.

What has happened since? The key events are the Lisbon and Edinburgh Summits in 1992, the Brussels Summit in 1993, and the inter-institutional agreement.

Lisbon Summit (June 1992)

The communiqué stated that the principle of subsidiarity was to be consciously considered for all new EC legislation, but also called for a re-examination of certain EC rules to adapt them to the principle. The subject was to be discussed again at the Edinburgh Summit (December 1992) and a report on the re-examination of existing EC rules was to be prepared for the December 1993 Summit.

In the period running up to the Edinburgh Summit some Member States submitted lists of Directives for repeal or 'repatriation'. Some became public and caused controversy. The Commission in October issued a communication – SEC (92)1990 – on the 'Principle of Subsidiarity' which did not list any Directives but raised the possibility of Member States supervising their own application of EC environmental legislation. This of course they do anyway although the Commission has a duty under Article 155 to ensure that Community rules are respected (see Chapter 14).

Edinburgh Summit (December 1992)

The Conclusions of the Presidency included a document on subsidiarity discussing: (a) basic principles; (b) guidelines (for determining whether a proposal for a Community measure conforms to Article 3b); (c) procedures and practices. The document concluded with 'examples of the review of pending proposals and existing legislation'. No proposals for possible withdrawal concerned the environment (unless the proposed zoos Directive can be called environmental).

Concerning existing legislation, the document had this to say:

> On the environment, the Commission intends to simplify, consolidate and update existing texts, particularly those on air and water, to take new knowledge and technical progress into account.

This was widely interpreted as the abandonment of the idea of 'repatriation'. What has happened is that the debate about subsidiarity has been used to argue that certain existing Directives should be improved.

Three guidelines were set out in the Edinburgh document to answer the question whether the Community should act in any particular case:

- issues having transnational aspects;
- actions by Member States that distort competition;
- action at Community level producing clear benefits by reason of scale or effects compared with action at national level (sometimes referred to as 'intensity' or 'proportionality').

The last of these is the most problematic because, more than the other two, it is a matter of judgement. Bathing water and drinking water hardly fall under the first two categories, but it can be argued that they do fall within the third.

The Edinburgh Summit also proposed an inter-institutional agreement on subsidiarity between the European Parliament, the Council and the Commission.

Brussels Summit (December 1993)

During 1993 an Anglo-French list of EC legislation was sent to the Commission but also became public. Among those dealing with other subjects it included for repeal or amendment the following environmental Directives (without saying which were to be repealed and which were to be amended):

- shellfish water (79/293);
- water for freshwater fish (78/659);
- drinking water (80/778);
- bathing water (76/160);
- air quality standards (80/779, 85/203, 82/884).

Before the Brussels Summit in December 1993 the Commission submitted a report on the adaptation of Community legislation to the Subsidiarity Principle – COM(93)545 – which went into some detail about the Commission's intentions for some of the water and air Directives and made it clear that Directives were only to be repealed once they had been replaced by others. The Council Conclusions, by contrast, were rather brief but noted with satisfaction that the Commission was withdrawing a number of proposals and suggesting the repeal of certain existing legislative acts and the simplification or recasting of others. It did not say which. The conclusions asked the Commission to submit formal proposals for adoption as speedily as possible.

The initiative therefore now rests with the Commission, but any proposal to amend the existing Directives will require a qualified majority in the Council and also must satisfy the Parliament whose powers have been increased by the Maastricht Treaty.

The Commission has made public some of its ideas for a revised drinking water Directive. Instead of appearing to be a complete statement of requirements for drinking water, as does the present Directive, the new Directive would set standards for some essential quality and health parameters, but would leave Member States free to add secondary parameters, for example the aesthetic parameters, if they see fit. Thus the Directive could give greater flexibility but essential standards would still be set at EC level.

The inter-institutional agreement – October 1993

On 25 October 1993 the Parliament, the Council and the Commission adopted an inter-institutional agreement 'on procedures for implementing the principle of subsidiarity'. This commits the Commission to include a justification under the principle of subsidiarity for any legislative proposal it makes and to draw up an annual report for the Parliament and the Council on compliance with the principle. The Parliament is to hold a public debate on that report.

Conclusions

Environmental policy has long been the subject of subsidiarity within Member States and has been the test bed for the development of the concept in the EC. All existing EC environmental legislation has been based upon it and although the debate about subsidiarity had not been particularly prominent before 1992, the idea has always been there. As a result, very few existing items concerning the environment are serious candidates for repeal. The end result of the great debate is likely to be the modification of a few environmental Directives and a general tendency for the EC's hand to be lighter rather than heavier in future. The inter-institutional agreement, with its requirements for an annual report from the Commission and an annual debate in the Parliament, will ensure that the issue of subsidiarity does not go away. We will continue to discuss whether the EC's hand is a heavy hand or a light one.

Paper presented in October 1993 and published in 1994

Developments since 1993

Subsidiarity is now defined in words that go beyond the words in the Maastricht Treaty by recognising that some actions are best taken at regional or local level. The choice of level within a Member State is of course entirely a matter for the Member States themselves. Article 5(3) of the consolidated version of the Treaty on European Union reads:

> the Union shall act only if and in so far as the objectives of the proposed action cannot be sufficiently achieved by the Member States, either at central level or at regional or local level, but can rather, by reason of the scale or effects of the proposed action, be better achieved at Union level.
>
> The institutions of the Union shall apply the principle of subsidiarity as laid down in the Protocol on the application of the principles of subsidiarity and proportionality. National Parliaments ensure compliance with the principle of subsidiarity in accordance with the procedure set out in the Protocol.

The procedure that enables the national Parliaments to exercise their important new role is remarkably little known. It is set out in the Protocol No. 2 to the Treaty into which the 1993 inter-institutional agreement, mentioned in the paper above, has now been subsumed. The Protocol also requires the Commission to issue annual reports on subsidiarity and it has done so ever since 1993.

If enough national Parliaments object they can issue what are called yellow and orange cards, following an analogy from the world of football. The Commission has to send draft legislation to national Parliaments and within eight weeks any national Parliament can send a reasoned opinion stating why it considers that it does not comply with the principle of subsidiarity. Each national Parliament has two votes, with the two chambers of a bi-cameral Parliament having one vote each. If there are more reasoned opinions than one-third of the possible total then the draft must be reviewed. This is known as 'showing a yellow card'. After reviewing the draft the Commission must give reasons for maintaining, amending or withdrawing it.

If national Parliaments representing a simple majority challenge draft legislation on grounds of subsidiarity, then it is said that an orange card has been shown. If after a review the Commission wishes to maintain the draft, it must issue a reasoned opinion which, together with the reasoned opinions of the national Parliaments, is then sent to the Council and Parliament. The Commission's proposal can then, in certain circumstances, be rejected by 55 per cent of members of the Council or by a majority in the Parliament.

Only two yellow, and no orange, cards have so far been issued – neither concerned with the environment – and the system has been criticised for being ineffective. One reason is that the eight-week time limit is very tight. Its merit is twofold. It provides pressure on the Commission to take subsidiarity seriously when making proposals, and it gives national Parliaments a motive for scrutinising proposed EU legislation. The fact that the number of reasoned opinions have been steadily rising suggests that national Parliaments are steadily becoming more engaged with EU policy.

The prediction made in the paper above – that the end result of the subsidiarity debate of the early 1990s would be that a few environmental Directives would be modified but that none would be repatriated – has been borne out by events. The Deputy Director-General for the Environment, at the conference at which the above paper was given, spoke of modifying some existing legislation 'from the point of view of up-dating them so far as technical progress is concerned, and making them less prescriptive' (Garvey 1994). He gave only three examples: the proposed 'packaging' and 'landfill' Directives (see Chapter 6) and the existing 'drinking water' Directive which was then being revised. The new 'drinking water' Directive was certainly less prescriptive than the one it replaced, but as described in Chapter 6 the Parliament was to kill the first attempt at a 'landfill' Directive and insisted that the one later adopted was prescriptive enough to achieve its objectives. There are rather few instances of environmental legislation adopted since the Maastricht Treaty having been modified during the process of negotiation for reasons of subsidiarity. One example is the SAVE Directive 93/76, which was amended before being adopted to remove detailed standards for thermal insulation of buildings in different climatic zones, and its replacement with a requirement for Member States to establish their own programmes – see Chapter 9. Subsidiarity has also affected proposals for reducing energy use in transport and agriculture – see Chapter 15. As subsidiarity has been a consideration for EU environmental policy since its inception, the modification to the SAVE Directive might well have happened anyway without the heightened attention to the subject generated by the Maastricht Treaty. Subsidiarity will continue to apply pressure.

References

Dubrulle, M (1994) *Future European Environmental Policy and Subsidiarity*, Brussels: European Interuniversity Press.

Garvey, T (1994) The destiny of selected environmental legislation. In: Dubrulle, M, ed., *Future European Environmental Policy and Subsidiarity*, Brussels: European Interuniversity Press.

Haigh, N (1994) *The Environment as a Test Case for Subsidiarity*, Environmental Liability CS22 (subsequently also published in Dubrulle 1994).

13

THE PRECAUTIONARY PRINCIPLE

Taking precautions against uncertain risks is nothing new but its elevation into a principle under German influence has brought it to prominence as an important element of the making of environmental policy both nationally, internationally and in the EU. The 'precautionary principle', as the *Vorsorgeprinzip* is usually translated into English, was already being discussed in Germany in relation to environmental policy in the 1970s, but it did not play a particularly significant role there until the 1980s. It has been suggested that it became prominent only when the die-back of German forests became a matter of public concern. A justification was then needed for the extra measures to be taken against air pollution that went beyond those already in place, and accordingly the *Vorsorgeprinzip* was called in aid (von Moltke 1988). As we have seen in Chapter 4, the German Government had in 1982 changed its stance on reducing emissions to air that were implicated in acid rain and forest die-back, and began retrofitting more power stations with abatement equipment.

In 1986, the German Government published 'Guidelines' for *Vorsorge* (precaution) discussing the subject in some detail – see Box 13.1. The principle then gathered momentum in international circles and with the signing of the Maastricht Treaty in 1992 it joined several others on which EU environmental policy is based (the principles that preventive action should be taken, that damage should be rectified at source, and the polluter pays principle. Although not stated as a 'principle', the Treaty also says that EU environmental policy 'shall aim for a high level of protection taking into account the diversity of situations in the various regions'.

The Treaty does not define the principle, but some indication of what was intended had already appeared in several international pronouncements. Although 'precaution' had previously been stated as a reason for taking action on specific international problems – protecting the ozone layer and the North Sea – the first general declaration was made at a conference of Ministers from 34 countries organised at Bergen in 1990 by the Norwegian Government and the United Nations Economic Commission for Europe (UNECE). This was part of the follow-up to the Brundtland Report and was intended to prepare for the UN Conference on Environment and Development (UNCED) to be held in Rio de Janeiro two years later. The 'Bergen Declaration' included the following wording:

> In order to achieve sustainable development, policies must be based on the precautionary principle. Environmental measures must anticipate, prevent and attack the causes of environmental degradation. Where there are threats of serious or irreversible damage, lack of full scientific certainty should not be used as a reason for postponing measures to prevent environmental degradation.

Principle 15 of the 1992 UNCED 'Rio Declaration' repeats some of the wording of the Bergen Declaration, but with some differences:

> In order to protect the environment the precautionary approach shall be widely applied by States according to their capabilities. Where there are threats of serious or irreversible damage, lack of full scientific certainty shall not be used as a reason for postponing cost-effective measures to prevent environmental degradation.

The Rio Declaration uses the words 'precautionary approach' rather than 'principle' as used at Bergen, suggesting reservations among some countries about what might be implied by a 'principle'. In the USA, in particular, the 'precautionary principle' has not been generally welcomed even though many examples of precautionary measures have been adopted there. These reservations are probably culturally conditioned, with Anglo-Saxon countries, which pride themselves on their pragmatism, shying away from the idea of elevating the common-sense notion of taking precaution into a principle that might involve unpredictable legal constraints. In his analysis of the *Vorsorgeprinzip* in Germany, von Moltke anticipated these reservations when he contrasted the detailed and explicit nature of German law – for which he gave reasons[1] – with the different form of American legalism which for other reasons is 'characterised by its pragmatism and its avoidance of general principles wherever possible' (von Moltke 1988). In Britain reservations were also to be found, and although a Minister stated in Parliament in 1988 – four years before it became enshrined in the EU Treaty – that the Government accepted the precautionary principle, many official British statements continued to refer to a precautionary 'approach' or 'action' or 'measures' (Haigh 1994).

As argued below, the precautionary principle is not a precise rule, and when applying it account has to be taken of other principles such as the principle of proportionality, so that the distinction between a 'precautionary approach' and the 'precautionary principle' is in practice more symbolic than real, however strong are the feelings are of those who prefer to avoid referring to principles.[2]

The introduction of the principle into the EU Treaty was followed by an extraordinary outpouring of literature on the subject, which continues to this day. Some of it discusses whether it is consistent with science or 'good science', and some even whether it is a principle of science. There are many aspects of the principle which are not self-evident, such as whether it is a principle of law (and if so, what that implies), when it can be used, by whom, and what level of risk makes its use appropriate. Some of this literature has generated confusion, some perhaps deliberately so since those with vested interests are often inimical to it.

It helps to reduce confusion if three key aspects are kept in mind:

- first, and most importantly, it is a principle of policy-making;
- second, it must be distinguished from the long-established principle of prevention;[3] and
- third, it must take account of the principle of proportionality.

Scientific knowledge is of course essential since those making and administering policy have to know as much as possible about how serious or irreversible is the risk, and the costs of action and lack of action. But the making of a policy decision is certainly not a scientific process (see Chapter 10), however much it depends on science, so that much of the literature on whether the principle is scientific is beside the point.

These three aspects were elaborated upon in a lecture I gave at an international seminar on environmental law held at University College London in 1997. The lecture was never written down but the notes distributed to participants – many of them American – survive.

The precautionary principle at work in the EU – notes for a lecture given in London, May 1997

A principle of policy-making

As it originally evolved in Germany the precautionary principle is a principle that relates to the making and administration of public policy (see Box 13.1). There has been heated discussion as to whether it is also a principle of science (Wynne and Meyer 1993)[4] and a continuing (inconclusive?) discussion as to whether it is also a principle of law (Cameron 1994; Freestone and Hey 1996).

The distinction between precaution and prevention

The Treaty of Rome as amended at Maastricht confirms that the principle relates to policy-making. Article 130r(2) states that 'Community policy on the environment ... shall be based on the precautionary principle and on the principle that preventative action should be taken'. Thus the precautionary principle must be distinguished from the principle that preventative action shall be taken. The distinction is discussed in the German Government's Guidelines on *Vorsorge* of 1986. Prevention relates to known harm. Precaution relates to risk of harm, i.e. uncertainty. The difficulty is that there is often uncertainty as to when scientific knowledge is sufficient to remove uncertainty. The initial decisions to restrict use of CFCs were taken in the absence of proof that the ozone layer was being damaged and were therefore precautionary. Now that there is a scientific consensus that CFCs damage the ozone layer it is more appropriate to talk of preventative action, but in between there was a period of uncertainty as to which principle was relevant. Much discussion of precaution is in fact about prevention.

Need to take account of the principle of proportionality

Application of the precautionary principle must take account of the principle of proportionality (since otherwise it is unworkable). This is made clear in the German Guidelines even though public authorities are anyway bound by the Basic Law to respect the principles of proportionality and the prohibition on excessive actions. Since these principles are not embedded in a written British constitution it is important here to emphasise them. Most statements of the precautionary principle do indeed say the principle is only to be applied where the risk is sufficiently serious. The policy-maker or administrator therefore has to exercise judgement. The principle is just that: it is not a precise rule.

Examples of application in the EC

The precautionary principle was applied in the EC before it was written into the Treaty, so no dramatic change should be expected as a result of Treaty change. Some examples of its use will be given. It is worth noting that it was first explicitly used in the EC in 1980 in relation to CFCs and indeed it was in relation to the ozone layer that it was first introduced into an international convention (Haigh 1994). (Both Freestone and Cameron incorrectly state that it appeared first internationally in relation to marine protection.)

Benefit and disadvantage of the principle

The principle has been criticised as vague and has certainly generated a great deal of obscurantist writing. Its benefit is that it provides a firm answer to those who think no action should be taken without conclusive proof of harm. Its possible disadvantage is that it may too readily give encouragement to those with axes to grind, e.g. a total ban on chlorine; investment in rocket-powered H bombs to fend off asteroids. Judgement remains essential.

28 April 1997

BOX 13.1 THE *VORSORGEPRINZIP* (PRECAUTIONARY PRINCIPLE)

The Federal German Government described the *Vorsorgeprinzip* in 1976 in the following terms:

> Environmental policy is not fully accomplished by warding off imminent hazards and the elimination of damage which has occurred. Precautionary environmental policy requires furthermore that natural resources are protected and that demands on them are made with care.

In 1986 the Federal Government published Guidelines for *Vorsorge.* These define environmental *Vorsorge* in the following terms:

> The Federal Government bases its environmental policy on a broad concept of *Vorsorge.* As a principle for political action, environmental *Vorsorge* comprises all actions which serve:
>
> - the protection against specific environmental hazards;
> - the avoidance or reduction of risks to the environment before specific environmental hazards are encountered; and
> - in a future perspective the management of our future environment, in particular the protection and the development of the natural foundations of life.

The application of the *Vorsorgeprinzip* has to take account of other principles, in particular the general principles of West German administrative law:

- *Verhältnismäßigkeit* – the principle of 'proportionality' of administrative action; and
- *Übermaßverbot* – the prohibition of excessive actions.

These together require actions to be necessary and appropriate to the achievement of legally prescribed goals.

Hence, the Guidelines state that:

> The determination of measures for *Vorsorge* requires a balancing which takes into account on the one hand the economic and other effort involved and on the other the achievable maintenance and improvement of environmental quality.

In practice in West Germany, *Vorsorge* is often taken to be synonymous with emission standards in accordance with the state of technology as, for example, in the air pollution legislation which states:

> Installations subject to authorisation are to be constructed and operated in such as a manner that ... precaution is taken against damaging environmental effects, in particular by means of measures for the control of emissions in accordance with the state of technology.

The *Vorsorgeprinzip* exists in the context of other principles which are also relevant to environmental policy, such as the obligation of the state to deal with identifiable risks (*Gefahrenabwehr*), the polluter pays principle or a principle of public responsibility (*Gemeinlastprinzip*). The last is sometimes enunciated to justify public expenditure on environmental protection, particularly when polluters cannot be identified or for speed of action.

Taken from the 12th Report of the Royal Commission on Environmental Pollution highlighting points in the IEEP report of 1987 written by Konrad von Moltke, reprinted as Appendix 3 of the RCEP report (von Moltke 1988).

Developments since 1997

The Commission issued a Communication on the Precautionary Principle in 2000 (COM(2000)1) with a summary that starts by conceding that it is much disputed: 'The issue of when and how to use the precautionary principle, both within the EU and internationally, is giving rise to much debate, and to mixed, and sometimes contradictory views.' The Communication asserts that the principle has 'become a full-fledged principle of international law' and that its use by the EU is allowable under World Trade Organization (WTO) rules. The Communication wisely does not claim to be the final word and it was quickly followed by a Council Resolution[5] providing guidelines for the Commission but generally endorsing its Communication.

The second sentence confirms that it relates to the making of public policy: 'decision-makers are constantly faced with the dilemma of balancing the freedom and rights of individuals, industry and organisations with the need to reduce the risk of adverse effects to the environment, human, animal or plant health'.

The summary goes on to emphasise the role of science:

> The precautionary principle, which is essentially used by decision-makers in the management of risk, should not be confused with the element of caution that scientists apply in their assessment of scientific data. Recourse to the precautionary principle presupposes that potentially dangerous effects deriving from a phenomenon, product or process have been identified, and that scientific evaluation does not allow the risk to be determined with sufficient certainty. The implementation of an approach based on the precautionary principle should start with a scientific evaluation, as complete as possible, and where possible, identifying at each stage the degree of scientific uncertainty.

That sentence just quoted, together with the several further paragraphs that discuss the role of science, should provide a firm answer to those who contend that a 'precaution-based approach' is to be contrasted with a 'science-based approach', with the former being practised in Europe and the latter in the USA. There is certainly conflict but it should not be about the reliance on science, but about how decisions are made and their outcome. Industrialists whose interests are adversely affected by decisions that rely on the precautionary principle can be expected to protest, and they do. Some go further and attack to principle itself.

In its 2002 judgement in two cases that challenged a ban on the use of certain antibiotics in animal feed, the European Court of First Instance laid stress on the essential role of scientists, but clarified the difference between their role and the role of policy-makers. In supporting the ban it stated:

> Before taking any preventive measure, a public authority must therefore carry out a risk assessment, which involves two components: a scientific component, i.e. as thorough a scientific assessment as possible account being taken of how urgent the matter is, and a political component ('risk management') in the context of which the public authority must decide on the measure it deems appropriate given the degree of risk set by it.
>
> *(ECJ Cases T-13/99 and T-70/99)*

In other words the risk management component of a risk assessment, while dependent on science, is not a wholly scientific process. It is because so many find that difficult to grasp that the charge is made that the principle is unscientific.

The difference between precaution and prevention is not expressly discussed in the Communication, but is implied in this sentence:

> Whether or not to invoke the Precautionary Principle is a decision exercised where scientific information is insufficient, inconclusive, or uncertain and where there are indications that the possible effects on the environment, or human, animal or plant health may be potentially dangerous, and inconsistent with the chosen level of protection.

The implication is that where scientific information is indeed sufficient the precautionary principle need not be invoked. Measures can instead be taken to prevent damage by invoking the long-established principle of prevention: it is better than cure.

Proportionality is emphasised. Measures 'must not be disproportionate to the desired level of protection and must not aim at zero risk, something which rarely exists'. 'In some cases a total ban may not be a proportional response to a potential risk. In other cases it may be the sole possible response to a potential risk.'

Examples of the principle in use

Disappointingly, the Communication does not provide examples of the use of the precautionary principle in the EU apart from the Court cases on animal feed. It should not have been difficult to provide a comprehensive list as there were surprisingly few examples at that time. A list today would help to ground the continuing debate more firmly on how the principle is being applied in practice. No claim to completeness is made for the examples given here.

The principle has been applied in EU legislation to ban or restrict the individual chemicals listed below, but it has also been embedded in several items of EU legislation placing an obligation on Member States to apply it when making their own decisions, or on the Commission when making proposals. These include the IPPC Directive 96/61 (when determining the Best Available Techniques for reducing emissions – see Chapter 8)[6], the 'water framework' Directive 2000/60 (when dealing with priority hazardous substances – see Chapter 5); the 'habitats' Directive 92/43 (when making assessments under Article 6 of plans that have an effect on Special Areas of Conservation)[7]; the 'deliberate release of GMOs' Directive 2001/18; and Regulation 178/2002 establishing the European Food Safety Authority. It is also embedded in the Common Fisheries Policy (in Recital 10 of Regulation 1380/2013).

Two examples of its use in the EU result from international meetings or conventions. The Third North Sea Conference invoked the precautionary principle in calling for an end to the dumping of sewage sludge and this was made binding by the 'urban waste water treatment' Directive 91/271 (see Chapter 5). The Framework Convention on Climate Change signed at Rio in 1992 states that 'the parties should take precautionary measures to anticipate, prevent or minimise the causes of climate change and mitigate its adverse effects'. To fulfil obligations under the Convention the EU adopted the 'monitoring mechanism' Decision 93/389 which, as explained in Chapter 9, required each Member State to limit its emission of greenhouse gases. Since in 1993 there was uncertainty about whether climate change was of human origin, this is another instance of reliance on the precautionary principle.

The first explicit use of the principle by the EU occurred many years before it entered the Treaty and concerns the protection of the ozone layer (see Chapter 2). The EU is a party to the 1985 Vienna Convention for the protection of the ozone layer which refers to precautionary measures 'which have already been taken at the national and international levels'. These will have included the EU Decision 80/372 of 1980, which expressly stated that the cap on production of chlorofluorocarbons (CFCs) was a precautionary measure. The hypothesis had been advanced in 1974 that the release of CFCs could result in the depletion of the stratospheric ozone layer. This would increase ultraviolet radiation reaching the earth resulting in more skin cancers and also affecting vegetation and ecosystems, for example oceanic plankton which forms a fundamental part of the food chain. There was therefore no doubt that the risks were potentially very serious, but as there was then no proof that the hypothesis was sound, it was appropriate to describe the action taken as precautionary. In 1987 the Montreal Protocol went on to require cuts in CFC production, again as a precautionary measure. However, Joe Farman, who published his discovery of the ozone hole in 1985, maintains that the Protocol was not precautionary, unlike EU Decision 80/372 (Farman 2001). Certainly very soon after 1987, if not before, a scientific consensus emerged that CFCs were the cause of the 'hole'. The uncertainty about when there was a consensus goes to show that it is sometimes difficult to distinguish 'precaution' from 'prevention', not that the distinction has significance once a decision to act has been taken. The distinction is important only when action is being contemplated: if there is proof that some activity is sufficiently dangerous, a public authority has an obligation to act; whereas if proof is incomplete, the precautionary principle provides the public authority with the discretion to act. Industrialists would thus have more difficulty in successfully challenging a decision that adversely affects them, on the grounds that harm has not been scientifically established, if that decision is based on an established principle.

It should be no surprise that the principle has most often been applied at EU level to individual chemical substances, given that such decisions have to be taken at EU level to preserve the integrity of the single market (see Chapter 7). Also, as von Moltke noted in his analysis, the German Guidelines of 1986 'focus on the introduction of substances into the environment whether it occurs in air, water, or soil' (von Moltke 1988, section 3.2). The 'REACH' Regulation 1907/2006 on chemicals specifically refers to the precautionary principle in Article 1(3) and in its Preamble. Regulation 1107/2009 on plant protection products (pesticides) and Regulation 528/2012 on biocides also refer to the principle.

Some of the examples listed below are more fully discussed in a report prepared for the Commission (Milieu Ltd 2011).

- Hormones in beef. By Directive 81/602 the EU restricted the use of hormones to promote growth in animals, and the current restrictions are set in Directive 96/22 which replaced several earlier Directives. The ban provoked a long-running dispute with the USA which led to WTO dispute-resolution proceedings. During these cases the EU maintained that it was relying on the precautionary principle to justify its restrictions.
- Phthalates are additives used to make rigid plastics more flexible. Some have been shown to be reproductive toxicants and have endocrine-disrupting properties. Despite lack of evidence of sufficient leaching from plastics to cause harm, their use in toys and babies' teething rings was made the subject of a temporary ban in 1999 by Decision 1999/815. This was made permanent in 2005 by Directive 2005/84.
- Pentabrominated diphenyl ethers are flame-retardants which have been shown to be persistent bioaccumulative and toxic (PBT) substances. Their use was banned in 2003 by Directive 2003/11.
- Fenamirol is a fungicide which may have endocrine-disrupting properties. Its use was restricted under Directive 2006/134. This was unsuccessfully challenged – Case C-77/09.
- Bisphenol A (BPA) in babies' bottles was banned in 2011 by Directive 2011/8 because of concern that they have endocrine-disrupting properties.

- Antibiotics in animal feed. Regulation 2821/98 banned the use of four antibiotics on the grounds that bacterial resistance might be transferred to humans. The Court rejected a challenge (see p. 156).
- Neonicotinoid pesticides. In order to protect bees three pesticides have been banned temporarily under Regulation 485/2013.

A special case of the precautionary principle concerning chemicals is the decisions taken not by a public authority, but by a manufacturer as a result of EU legislation. Directive 79/831 (now replaced by REACH) required all new chemicals to be tested before being marketed (see Chapter 7). Some chemicals manufacturers would invent and develop a new chemical and either register it in the Chemicals Abstract Service or apply for a patent, so the existence of the chemical would eventually be known even if it was never marketed. It was thus possible for a competitor to enquire why, and anecdotal evidence suggests that one reason why some particular chemicals were never marketed is that preliminary tests showed that they might be very dangerous and not worth the costs of development. This can be regarded as the manufacturer invoking the precautionary principle, without the authorities ever knowing. A similar case occurs under REACH. Chemicals that have been identified as of very high concern are placed on the Authorisation List and a date is set when they are to be banned unless authorised. As discussed in Chapter 7, for some of the chemicals listed no application has so far been made for authorisation and the manufacturer has removed the chemical from the market, either because an alternative has become available or because the procedure to obtain authorisation was judged to be too uncertain.

The fact that this chapter has been insistent that the precautionary principle was enunciated as a principle for the making of public policy does not mean that it cannot be invoked by others. Every time a private individual decides to cross a road only at an authorised crossing rather than risk unpredictable fast-moving traffic, the principle is being invoked. The cases above are slightly different since the chemical manufacturer has anticipated what the decision of the public authority might be under EU legislation. Both the individual and the chemical manufacturer are making policy decisions, not scientific decisions.

The European Environment Agency has published two reports titled *Late Lessons from Early Warnings* (EEA 2001, 2013), the first subtitled *The Precautionary Principle 1896–2000* and the second *Science, Precaution, Innovation*. The two reports between them discuss 34 case studies covering such subjects as asbestos, radiation, benzene, polychlorinated biphenyls, CFCs, antibiotics in animal feed, tributyltin, hormones in beef, lead in petrol, bisphenol A, climate change, neonicotinoid pesticides and nanotechnology. The lead editor, David Gee, has said that in only very few of these cases has the precautionary principle been applied at EU level, though it has been applied by Member States in some instances.

What conclusions can be drawn from the EU's involvement with the precautionary principle? The first point to make is that it was one Member State – Germany – that developed the principle and that, by first championing it at various international fora, persuaded the other Member States to introduce it into the Treaty. Given that it has now matured and is widely accepted as an important innovation, it is a surprise that it has been invoked rather rarely at EU level, and that it continues to be controversial. It is no surprise that those whose interests are affected do not like it, but what is a surprise is that the very idea of the principle continues to come under attack. To those attacking it one can pose this question: should the restrictions on CFCs that began in the late 1970s and early 1980s really have been delayed till 1987 when a scientific consensus emerged as to their seriously damaging effects?

Notes

1 He attributes this partly to Germany's federal structure and partly as a reaction to the experience of National Socialism. He does raise an eyebrow over the need for the 'principle of cooperation' – also discussed in the Guidelines – which 'most countries would consider a natural prerequisite of a functioning political system'.
2 Some people insist that they accept a precautionary approach in risk assessment, i.e. safety factors to extrapolate from rats to humans, etc., and that this does not imply an acceptance of the precautionary principle. The Commission's Communication deals with this point by noting that 'the element of caution that scientists apply to their assessment of scientific data' should not be confused with the precautionary principle used by decision-makers in their management of risk.
3 The principle of prevention was one of eleven principles set out in the EU's First Action Programme on the Environment adopted in 1973. It was introduced into the Treaty by the Single European Act of 1987. The precautionary principle was added by the Treaty of Maastricht of 1992.
4 The article by Brian Wynne and Sue Meyer in *New Scientist* (Wynne and Meyer 1993), 'How science fails the environment', prompted an extensive correspondence which was published as a *New Scientist Newsletter*, much of it about whether the precautionary principle was scientific.
5 Adopted at the European Council meeting in Nice in December 2000. In places it is more nuanced than the Commission's Communication, as when it says that the principle is only 'gradually asserting itself as a principle of international law'. It confirms that while scientists must talk to policy-makers, it is not scientists who take policy decisions. It expresses this in tactfully opaque bureaucratic language: 'affirms that those responsible for scientific assessment of risk must be functionally separate from those responsible for risk management, albeit with ongoing exchange between them'.
6 The obligation was dropped when the IPPC Directive was subsumed into the 'industrial emissions' Directive 2010/75. Was this inadvertent or deliberate?
7 In Case C-127/02 – known as the Waddensee case – the ECJ held that it was unacceptable to fail to undertake an appropriate assessment of possible impacts on a Natura 2000 site on the basis that significant effects are uncertain.

References

Cameron, J (1994) The status of the precautionary principle in international law. In: O'Riordan, T and Cameron, J, eds, *Interpreting the Precautionary Principle*, London: Earthscan.

EEA (2001) *Late Lessons from Early Warnings: The Precautionary Principle 1896–2000*, Environmental Issue Report No. 22, Luxembourg: Office for Official Publications of the EC/European Environment Agency.

EEA (2013) *Late Lessons from Early Warnings: Science, Precaution, Innovation*, Report No. 1/2013, Luxembourg: Office for Official Publications of the EC/European Environment Agency.

Farman, J (2001) Halocarbons, the ozone layer and the precautionary principle. In: *Late Lessons from Early Warnings: The Precautionary Principle 1896–2000*, Environmental Issue Report No. 22, Luxembourg: Office for Official Publications of the EC/European Environment Agency.

Freestone, D and Hey, E, eds (1996) *The Precautionary Principle and International Law*, The Hague: Kluwer.

Haigh, N (1994) The introduction of the precautionary principle into the UK. In: O'Riordan, T and Cameron, J, eds, *Interpreting the Precautionary Principle*, London: Earthscan.

Milieu Ltd (2011) *Considerations on the Application of the Precautionary Principle in the Chemicals Sector*, Brussels: Milieu Ltd.

von Moltke, K (1988) The *Vorsorgeprinzip* in West German environment policy. In: *Royal Commission on Environmental Pollution 12th Report: Best Practicable Environmental Option*, Cm 310 (previously published as an IEEP pamphlet in 1987).

Wynne, B and Meyer, S (1993) How science fails the environment, *New Scientist*, 5 June.

14

MAKING THE LEGISLATION WORK

There was no political discussion about implementation of EU environmental policy during its first decade, and it took a public scare about the disappearance of drums of toxic waste to raise the temperature. The European Parliament reacted by appointing a Committee of Inquiry, and in 1984 it censured the Commission for its failure properly to perform its role of guardian of the Treaties. Since then the need for better oversight of implementation has become a constant refrain. It has featured in all subsequent Environment Action Programmes including the current Seventh Programme of 2013.

In 1996, the Commission issued its first Communication on the subject. To prepare for this the Parliament, jointly with the Commission, organised a Hearing, and I was invited to write the background paper for the participants (Haigh 1997). The message was simple: the subject is complicated and has many aspects. No one mechanism will ensure effective implementation. The main duty to implement rests, of course, with the Member States, but the Commission as guardian of the Treaties has to ensure that the resulting legislation is properly applied and since it cannot do everything itself it must depend on others. Although the paper is twenty years old it sets out, in a way not readily to be found elsewhere, why the subject remains high on the agenda (see Chapter 15).

In presenting my paper at the Hearing, I compared the origin of a Directive to the birth of a child. Somehow a Directive is first conceived. Then there is a period of gestation, often quite lengthy, until at a Council meeting, and to mounting excitement, the Directive is given a final push and enters the world to the joy and applause of the parents (Council, Commission). There the analogy stops. At that point the parents lose interest and turn their minds to the next Directive. In contrast, parents of real children know only too well that a birth is only the beginning of a very long road to adulthood. The child cannot survive on its own, but needs constant attention from its parents and from many others who contribute to its care and education.[1]

My interest in the subject had started in 1980 when, as my first project at the Institute for European Environmental Policy (IEEP), I studied the impact of EU environmental Directives on the UK. This involved reading all the relevant British legislation and related administrative documents, and then interviewing the responsible authorities and as many of those affected as I could reach. This led me to draw a distinction between formal implementation on one hand (what lawyers call the 'transposition' of EU law into national law) and, on the other hand, the effects in practice or 'practical implementation' (which is not always quite the same as what lawyers call 'application'). I realised that studying practical implementation was quite different from studying transposition. In 1982/83 IEEP presented the preliminary results of this work – later published in book form (Haigh 1984) – to Commission officials, who up till then had not paid attention to 'practical implementation' but had largely concentrated on checking 'transposition'. As a result the Commission asked IEEP to conduct a number of comparative reports on how different Member States were implementing certain Directives in practice. From this we came to appreciate the importance of understanding the different administrative structures in the Member States which result in Directives being implemented quite differently. The 1984 book identified 'implementation' as one of four issues that would dominate the future of EU environmental policy.

Effective environmental protection – challenges for the implementation of EC law – background paper for invited experts and participants presented at a Hearing at the European Parliament, Brussels, May 1996

Introduction

EC environmental legislation began in the 1970s but its implementation only became a matter for political attention during the 1980s. Even now the Community Institutions devote a much higher priority to the adoption of new legislation than they do to its implementation. Yet unless EC legislation is properly implemented it will fall into disrespect and the environment will not receive the protection that the EC legislators intended and that the public expects.

The first three Action Programmes on the environment of 1973, 1977 and 1983 did not recognise the importance of implementation, and it took a well publicised scare to draw attention to the subject. When some drums of hazardous waste, thought to contain dioxin from Seveso in Italy, disappeared and were subsequently discovered in France, the European Parliament promptly established a Committee of Inquiry (the Pruvot Committee). This reported on 'the odyssey of the drums' as result of which the Parliament adopted a Resolution (OJC 127/67 14.5.84) which censured the Commission 'for having failed to perform fully and properly its role of guardian of the Treaties' and 'for its failure to take the necessary measures vis-a-vis the Member States with regard to the implementation and application of the Directive' (Directive 78/319 on toxic waste). As a result, implementation began to be recognised as an important subject. The Commission responded by increasing the number of staff in DG XI dealing with implementation and became more vigorous in bringing cases before the Court of Justice. Also in the same year the Commission published the first of its annual reports to the Parliament on the application of Community law – COM(84)181 – giving some statistics on infringement proceedings and drawing some conclusions.

Shortly afterwards the Council, in adopting the Fourth Action Programme on the Environment in 1987, underlined 'the particular importance it attaches to the implementation of Community legislation' (OJC 328/2 7.12.87), and the Parliament identified several factors leading to inadequate implementation in a Resolution on implementation of water legislation (OJC 94/155 11.4.88).

The political debate about how implementation was to be taken more seriously had thus already begun when, at the highest EC level, an even stronger commitment was made by the European Council (the Heads of State and Government) when it adopted the 'Declaration on the Environmental Imperative' at Dublin in June 1990. This stated:

> Community environmental legislation will only be effective if it is fully implemented and enforced by Member States. We therefore renew our commitment in this respect. To ensure transparency, comparability of effort and full information for the public, we invite the Commission to conduct regular reviews and publish detailed reports on its findings. There should also be periodic evaluations of existing Directives to ensure that they are adapted to scientific and technical progress and to resolve persistent difficulties in implementation; such reviews should not, of course, lead to a reduced standard of environmental protection in any case.
>
> *(Bulletin of the EC 1990)*

It is against this background of concern and commitment that the Commission is now preparing a communication on implementation, and that this Hearing is being held. As the Dublin Declaration emphasised, the subject of implementation requires transparency. It is also now recognised as having many different aspects, with different actions being required by different actors including the Community Institutions, the Member States and the authorities within them, and all those affected by the legislation including the general public.

This background paper is an attempt to provide a brief overview of the many different aspects of the implementation of Community environmental law to provide a frame for the more detailed discussions at the Hearing.

Aspects of implementation

To regard implementation of EC law as a process which starts only after legislation is adopted is to omit an essential aspect. Ambiguous or poorly drafted legislation often leads to subsequent problems and therefore the process of adopting the legislation has to be regarded as the **first aspect** of implementation. Once adopted, Directives then have to be transposed into national law and administrative provisions. While EC Regulations are directly applicable law and do not themselves need to be transposed, they often require national legislation to supplement them, e.g. appointment of competent authorities and penalties for non-compliance. Transposition into national law can therefore be regarded as the **second aspect** of implementation.

But the main purpose of a Directive is not to develop national legislation, but is to achieve certain results. The steps taken in the Member States to achieve those results using national law and administrative provisions can be regarded as a **third aspect** that can be called practical implementation. Practical implementation may involve many steps including strengthening competent authorities, drawing up plans, following procedures, meeting standards, designating areas, providing information, and granting authorisations. It may involve investments – often very large – by the public and private sectors, perhaps resulting in new products and processes. It will involve monitoring and reporting. The practical implementation of some Directives, e.g. relating to a product, may affect rather few manufacturers and may be relatively simple and therefore easy to evaluate. Other Directives may involve hundreds of actions per year in a Member State, any one of which may not fully comply with the Directive so that assessment and comparisons between countries is difficult. It is important therefore not to draw general conclusions from a limited experience of only a few examples of implementation. Some Directives have indeed been successfully implemented often with little delay, while others are problematic and have involved long delays.

A **fourth aspect** of implementation is enforcement under the processes of national and Community law.

These different aspects sometimes overlap and some can be subdivided into yet further aspects. For example the appointment of a competent authority in an item of national legislation is part of transposition, but if a new authority has to be created or an existing one expanded with new staff and extra money, then that can be regarded as practical implementation. Monitoring is an aspect of enforcement, but if the details are prescribed in a Directive then it is also practical implementation. If a Directive requires a report evaluating a Directive after several years, then that is practical implementation, but if an evaluation is carried out in the absence of a requirement to do so then that can be regarded as a **fifth aspect** of implementation.

All these aspects: drafting, transposition, practical implementation, enforcement and evaluation are described below.

Drafting and adopting EC legislation

While the responsibility for proposing a Directive rests with the Commission, the finally adopted text is the result of a process of negotiations in the Council machinery, which increasingly also now involves the Parliament. The process often requires compromises as each Member State seeks to ensure that its own objectives are achieved, and the results show that problems of implementation have not always been adequately considered. Some examples can be given.

A number of Directives set numerical standards to be met by a certain date, a case in point being the drinking water Directive 80/778. The Directive, however, is silent about what should be done if the standard is not met by the deadline: should water supply be discontinued even if the risk to human health in doing so is greater than continuing to supply safe water which may not meet the standards? The Commission has now learnt from experience, and under the proposed new drinking water Directive, Member States will be able to grant time-limited derogations in certain circumstances subject to certain conditions. Another lesson can be drawn from the experience of the environmental impact assessment (EIA) Directive 85/337, which contained no transitional provisions dealing with developments authorised after the date set in the Directive but where the authorisation process had started before. The Commission interpreted the Directive one way and some Member States another way. A great deal of emotional energy was expended in many Member States before the Commission changed its interpretation, following an opinion of the Advocate General in a case before the Court of Justice relating to a power station in Germany (Case C431/92). Because of this ambiguity, expectations were falsely raised among some sections of the public and the Community's reputation has suffered as a result. A conclusion that can be drawn is that the Commission and Council must consider the need for transitional provisions before adopting a Directive.

A third lesson concerns reporting requirements which are important tools for evaluating what effect the Directive has had and how effectively it is being implemented. Some Directives include reporting requirements while others do not (see pp. 174–5). The drinking water Directive, for example, places no obligations on Member States to report on quality so it is not possible to compare the quality of drinking water across the Community or to know how well the Directive is being implemented. This lesson has been learnt in the proposed new drinking water Directive. In other Directives the reporting requirements are often inconsistent so that in 1991 Directive 91/692 was adopted to standardise and rationalise these requirements. However, the lesson is sometimes forgotten because the proposed new Directive on PCBs [polychlorinated biphenyls] (on which the Council adopted a common position in November 1995 – OJC 87/1 25.3.96) contains no reporting requirements.

One final point on drafting concerns the practice of agreeing ambiguous wording, perhaps as a way of securing agreement, and then recording explanatory statements in the unpublished Council minutes. An example concerns Directive 76/464 on the discharge of dangerous substances to water. The wording in the Directive that emissions limits are to be based on 'best technical means available' is qualified in the Council minutes by the statement that 'best technical means available is to take into account the economic availability of these means'. Not surprisingly, this practice of secretly modifying the language of legislation was criticised by the Parliament in 1984 (OJC 172 2.7.84) but it was not till October 1995 that the Council adopted a 'Code of Conduct on Public Access to the Minutes and Statements in the Minutes of the Council acting as a legislator', which should reduce the practice. It is too soon to assess whether the Code has been effective in curbing a practice that makes the assessment of implementation much more difficult, as well as offending against the principle that legislation should be publicly available.

One can conclude that, while the needs of implementation are being considered more seriously as new Directives are adopted, there is still much room for improvement. In particular the Community legislator (Commission/Council/Parliament) could ensure that before any Directive is finally adopted the needs of implementation are formally considered, including reporting requirements, transitional provisions and absence of ambiguity. A checklist might be helpful. Each Member State should also be prepared to give an assurance, at the time of adoption of a Directive, that it has begun to consider what is needed for implementation both by way of legislation, finance and capacity of its competent authorities. All too often this process does not start until the deadline for implementation is approaching. This would be to fulfil the call made by the Council in the Resolution adopting the Fifth Action Programme (OJC 138/3 17.5.93) which stressed that 'due regard should be given both at the stage when legislation is proposed and when it is adopted to the quality of the drafting of legislation, in particular in terms of the practicability of implementing and enforcing it'.

Transposition into national law

The requirement in Directives that Member States should communicate to the Commission the texts of the provisions of national law which they adopt to implement a Directive is more complicated than it sounds. Only rarely will one text be sufficient for implementing one Directive in a Member State.

Frequently different texts are required for different Articles of a Directive, particularly where they are the responsibility of different Ministries. In federal or regionalised Member States, where for constitutional reasons, or reasons of national administrative structure, competence for introducing the legislation rests with the regions, or is shared between the regions and the national government, different texts may be required for different parts of the national territory. In some Member States, such as Belgium, the national government has no authority over a region to ensure that legislation is adopted, even though Directives place obligations on Member States that are answerable to the Commission and the Court of Justice. There is also the problem that some Directives apply to some territories of Member States outside Europe (e.g. French overseas departments) or dependent territories (e.g. Netherlands Antilles, Gibraltar) while others do not. National laws (Acts of Parliament, Regulations) may also have to be supplemented by administrative documents (circulars, technical advice). It follows that the Commission will be sent many texts.

Member States frequently fail to communicate texts on time and this may be because of an oversight, because no texts are yet ready, or because some but not all texts are ready. There is also the problem that national laws are constantly evolving so that the texts sent to the Commission become out of date and have to be supplemented over time. The Commission's task in checking completeness and correctness of transposition, which should in theory be straightforward, is therefore difficult. It is not helped by the fact that communications between the Member States and the Commission is not always transparent since not all Member States make public the texts of their letters of transposition (sometimes called 'compliance notices' and sometimes set out in the form of 'tables of implementation' – '*tableaux de concordance*'). Thus the reasons for delay are often not easy to ascertain and are often not publicly explained. Since these documents provide a link between Community legislation (which is public) and national legislation (which is public) there should be no objection to publication of letters of transposition. Publication of such letters would enable the interested public (industrialists, academics, NGOs and other national governments) to check the adequacy of transposition and thus help the Commission by pointing to any deficiencies. Transparency would also create the pressure for improvement. Although the listing of implementation in the CELEX database is useful, CELEX does not indicate which Articles of a Directive are implemented by which items of national legislation.

One welcome development in recent years has been the requirement in EC legislation that national implementing legislation should contain a reference to the Directive that it is implementing, thus showing the link between national and EC legislation. Without this visible link the competent authorities applying national legislation sometimes did not to know that it is derived from EC legislation.

Practical implementation

A Directive is binding as to the ends to be achieved while leaving to national authorities the choice of form and method. The ends to be achieved may be that certain standards are to be met by a certain date or that certain procedures are to be followed for a given purpose. There are many different procedures laid down in environmental Directives, including:

- drawing up plans
- designating sensitive areas
- granting authorisations
- providing information
- making assessments
- conducting consultations
- restricting marketing and use
- monitoring
- preparing reports.

The steps taken to meet the standards and to fulfil the procedures all constitute practical implementation. Practical implementation is sometimes called 'application', which has been defined by the Commission as 'the incorporation of Community law by the competent authorities into individual decisions, for instance when issuing a permit or executing a plan or programme'. However, a Directive may also result in many decisions being taken by, for example, an industrialist, about which the competent authority may never be informed even though the industrialist's decision is taken as a result of implementing national legislation that transposes the requirements of a Directive. For example Directive 92/32 (known as the 'seventh amendment') on the notification of new chemicals requires a manufacturer to supply the competent authority with the results of studies for evaluating risk from the chemical before placing a new product on the market. The manufacturer may therefore discontinue development of a new chemical if their initial assessment suggests that it will be too dangerous to be worth manufacturing. The Directive will therefore have been effective in preventing use of a dangerous chemical, but no-one other than the manufacturer need know. Even if this is unknown, it must nevertheless be regarded as successful practical implementation of the precautionary principle embodied in the Directive.

Sometimes an existing competent authority will simply take on new tasks required by a Directive, but if a new authority has to be appointed, which may involve hiring and training new staff, then that too must be regarded as practical implementation.

There are now a very large number of items of EC legislation, and 485 are listed and discussed in the IEEP *Manual of Environmental Policy: The EC and Britain* (Farmer 2012). Not all these items are the responsibility of DG XI and many are minor amendments or are of a very detailed character (e.g. setting criteria for eco-labels). They can be divided into a number of categories according to practical steps that they require:

- product standards (e.g. lead in petrol, vehicle emissions, lawnmower noise)
- restrictions on production or marketing or use of substances (e.g. CFCs, asbestos, PCBs)
- emissions from stationary plant (e.g. certain plants require authorisations, and for some standards are set – e.g. sulphur dioxide, mercury)
- environmental quality standards (e.g. bathing water, air quality)
- designating areas (e.g. special protection areas for birds, nitrate-sensitive areas, waters for shellfish)
- notifications to competent authorities (e.g. new chemicals, dangerous installations)
- plans and programmes (e.g. pollution reduction programmes, emergency plans)
- assessments (e.g. environmental impact assessments of development projects, assessments of risks of chemicals).

Many Directives will fall under more than one category.

It follows that practical implementation is a very wide subject involving many different kinds of decisions and actions, some involving considerable investments over a long timescale. It is therefore extremely difficult (indeed impossible) to make any overall assessment of the adequacy of the practical implementation of all environmental Directives. They have to be considered one by one, and sometimes Article by Article. For some Directives practical implementation is fairly straightforward and easy to assess. This is particularly true of products.

For example petrol is manufactured and distributed by a limited number of enterprises in the EC. Because the petrol is a traded product, transposition of Directives setting standards for lead in petrol is usually achieved by means of national legislation rather than involving regions or local authorities. The national legislation can be drafted quickly and can be simple. Once the standard is set in national legislation it is easy for a competent authority to analyse petrol to ensure the standards are met, and for competitors to analyse each other's products. Any failure by a petrol distributor to meet the standards will therefore be quickly known. Transposition and practical implementation may have involved costs for refiners but the steps are straightforward. In this case it can also be demonstrated fairly easily that the quantity of lead emitted into the environment in Member States has dropped significantly following the implementation of the Directive and that the lead content of air has reduced. This Directive has therefore been successfully implemented, and as a general comment it can be said that Directives relating to the single market, such as product standards, are likely to be well implemented because competitors act as watchdogs to ensure that there is no distortion to competition.

Directives setting emission standards are more difficult to assess than product standards because they will involve many installations, each one of which will have to be individually authorised and each one of which will need to be monitored over time to ensure that the emissions standards set in a Directive are being met. Some Directives set standards for a class of installations which are limited in number (e.g. chloralkali plants) and it is therefore not too difficult to collect and compare the national authorisations (though each may be many pages long) and monitoring data, even though several competent authorities may be involved within one Member State. However, other Directives cover classes which include many installations (e.g. power stations, incinerators) so that the task of assessing practical implementation across the Community is much more difficult. Failure to meet the standards at any one plant on one occasion will constitute a formal failure of implementation by a Member State and could lead to an adverse judgement of the Court of Justice. A statement that a Member State is in breach of a Directive may give no indication whether the breach is widespread or is a rare event.

Directives which involve plans, or designating areas, or assessment are more complicated still. Designations will often be made by competent authorities at regional or local level with knowledge of local circumstances and there may be inconsistency in the criteria applied and disputes, for example over the boundaries of areas. Monitoring to ensure that environmental quality standards are met may need to be extensive and will need adequate numbers of trained staff.

Some Directives will involve more than one kind of competent authority, which may have to collaborate together to ensure practical implementation. For example the 'Seveso' Directive 82/501 on major accident hazards requires a manufacturer operating a hazardous installation to prepare a safety report and an 'on-site' emergency plan. The competent authority for supervising these is usually a technical inspectorate concerned with safety in factories ('labour inspectorate'). However, the Directive also requires the preparation of an 'off-site' emergency plan outside the installation and this is usually prepared by a local authority which is better placed to communicate with the public likely to be affected. But in order for the 'off-site' plan to be prepared, the local authority will need to have information from the manufacturer or competent authority responsible for the safety report and 'on-site' plan. Effective practical implementation therefore depends on good relationships between the competent authorities.

Some Directives have practical effects that touch activities that are not just technical but are embedded in the culture of a country. An example is the setting of the periods of the hunting season under the birds Directive, which have caused much controversy.

The Directive that has probably caused the greatest problems in practical implementation is Directive 85/337 on EIA [environmental impact assessment] of certain development projects. A large number of different kinds of development project are included within the scope of the Directive, so that different Government ministries and competent authorities are involved. Projects covered include industrial plant, transport projects (roads, railways, airports, etc.), installations for hazardous waste, and certain urban developments. For some projects an assessment is always required, while for others Member States have discretion to decide whether or not a project is likely to have a significant effect on the environment. Hundreds of assessments may be made under the Directive in a Member State each year, any one of which may not completely fulfil the requirements set in the Directive, particularly since assessments are often subjective. Since large development projects or even small ones in sensitive areas frequently cause controversy, the EIA Directive is subject to many complaints about failures of implementation.

While the variety of provisions within Directives is one cause of the difficulty in trying to assess their implementation, another is the variety of administrative structures within the Member States for implementing them. In some Member States there will be a single competent authority for some Directives, so that there is likely to be national consistency in implementation. In other Member States the constitution or administrative structure will require implementation to be carried out at the level of the regions and there may therefore be greater variation. In some Member States there may be specialised competent authorities, whereas in others the task may fall to local authorities who have many other tasks and priorities. In some Member States the competent authorities may be understaffed and inadequately trained.

Some Directives may require practices that have long been carried out in one Member State, while practices may be new in another. In theory therefore the Directive should be easy to implement in the first Member State, while the second may have to introduce completely new structures to ensure implementation. There is a paradox, however, in EC environmental policy that a Member State which already has a procedure before a Directive is adopted may have more difficulty in implementing it correctly than another Member State which starts freshly on the subject. This is because a competent authority that has already developed its traditions may not be convinced of the need to make what it may regard as unnecessary changes to existing practices in order fully to implement a Directive. Examples include implementation in the Netherlands of the EIA Directive, and in Germany of the Seveso Directive.

The above discussion suggests that the only effective way to consider practical implementation is to examine it Directive by Directive and Member State by Member State. Comparative reports can be prepared only with a full knowledge not just of the technical subject matter of a Directive, but also of the administrative structure and indeed the culture of the country implementing it. There is no short-cut to a study of practical implementation of all Directives in the Community as a whole. It follows that the very brief discussion in the annual reports on monitoring the application of Community law presented by the Commission are of limited value as a basis for a serious discussion that would help to improve practical implementation. The Dublin Declaration on the Environmental Imperative (see p. 164) of the European Council went further and called on the Commission to conduct regular reviews *and to publish detailed reports* on its findings (emphasis added). These need to relate not just to measurements but also to assessments of the adequacy of national administrative structures (numbers of staff, training, etc.).

Some basis for regular reviews and detailed reports is provided in some Directives in the form of monitoring obligations and obligations to evaluate and report. These are discussed below.

Financing

The costs of implementation – which are sometimes very large – are in principle a matter for the Member States and, following the 'polluter pays' principle, a matter for the polluter.

However, there are some sources of EC funds that contribute towards the implementation of EC environmental legislation. These include LIFE, the Structural Funds and the Cohesion Fund (for Spain, Portugal, Greece and Ireland).

The LIFE Regulation 1973/92 has provided modest funds for the improvement of administrative structures and environmental services (including the improvement of monitoring networks) and the promotion of environmental education, training and information (including exchange of experience on eco-management and auditing). For the period 1996–99 the Commission proposes to focus LIFE on four main areas, one of which is 'implementation of EU environmental policy'.

The Commission's twelfth annual report on 'monitoring the application of Community Law 1994' (OJC 254/44 29.9.95) has a section on Structural Funds in the chapter on the Environment. This notes that the integration of environmental concerns has been reinforced by new provisions in the Structural Funds Regulation

2081/93, which provides for the inclusion of an environmental appraisal in national and regional funding plans as well as retaining the previous provision concerning compliance with Community environmental law and policy of funded projects. The funds have been used for such matters as water treatment and waste management and thus for the implementation of EC environmental legislation.

Under the Cohesion Fund, large sums of money are available in Spain, Portugal, Greece and Ireland for environmental protection and transport projects. Projects funded have included sewage works to implement the Urban Waste Water Treatment Directive 91/271, and the provision of waste disposal facilities to implement the Waste Directive 75/442. The rules for the Cohesion Fund do not have the same environmental safeguards as do the Structural Funds so it is possible for sewage works built with the purpose of implementing one Directive to be located in an environmentally damaging way. Already the Court of Auditors has criticised EU spending in Spain and Greece because of environmental impacts. It is a further curiosity of the rules of the Cohesion Fund that each funded project shall cost at least 10 million ECU so that sewage works that are oversized are being built. This is not to the long-term benefit of either the Member States (because of maintenance costs) or of the environment. In its twelfth report on implementation (mentioned above) the Commission has commented that a regrettable deficiency of the spending of the Cohesion Fund is the failure in Member States to undertake forms of environmental planning required by EC legislation, for example the preparation of waste plans.

However, the Cohesion Fund Regulation does not require this as a condition of funding. Given that enlargement of the Community may require further Community funding to enable the countries of Central and Eastern Europe to implement EC environmental law, it is important that the lessons of the Cohesion Fund are learnt so that any new funding instruments have adequate environmental safeguards.

Monitoring

The word 'monitoring' is used in many different ways, from a narrow technical sense (e.g. using instruments to measure the presence of pollutants in emissions from industrial plants) to a broad sense which includes assessment or evaluations. In this section the word is used in the narrow sense of gathering data, without value judgements being made. Several Directives have monitoring obligations in this narrow sense to confirm that standards are met (sometimes called 'compliance monitoring') and these sometimes also lay down monitoring methods. These relate in particular to air and to water. The habitats Directive 92/43 requires Member States to undertake 'surveillance' of the conservation status of the habitats and species found on their territory.

Monitoring is an essential tool for implementation and Member States will have their own monitoring methods and frequencies which must at least conform with the requirements set out in Directives. Sometimes Directives require the results of monitoring to be published (see below) but even if there is no express requirement for publication it is possible for individuals to request access to this information under Directive 90/313 on freedom of access to information.

It is one of the tasks of the European Environment Agency (EEA) 'to help ensure that environmental data at European level are comparable and, if necessary, to encourage by appropriate means improved harmonisation of methods of measurement'.

Reporting/evaluation

Many Directives require Member States to submit reports. Some of these will just be factual (e.g. the quality of bathing waters), but some are in the nature of an evaluation of the implementation of a Directive after a number of years. Some Directives require the Commission to consider national reports and produce a Community-wide report. Some reports have to be published but some do not, and the record of publication so far is not good.

In November 1993 this Institute (IEEP) reviewed the state of reporting and found that of 64 items of environmental legislation requiring the Commission to produce reports on implementation, in the case of some 24 the Commission had produced no report at all, and for a further 16 reports had either been delayed (sometimes by years) or had been published at very infrequent intervals. The reasons for this included failures by the Member States to provide information – or information of the right sort – to the Commission, and cumbersome procedures and a lack of resources within the Commission itself. A further reason was the lack of pressure from the Council, with has given a higher priority to the adoption of new legislation.

The production of a Community-wide report involves many people and there is in effect a chain: regulated bodies report to competent authorities, competent authorities report to Member States, Member States report to the Commission, and the Commission reports to the Parliament and Council. There may be delays or deficiencies anywhere along this reporting chain which can therefore affect the final report. Some problems with the original reporting requirements have been recognised for some time, and in an attempt to standardise and rationalise implementation reports Directive 91/692 requires the Commission to publish reports on groups of Directives (air, water, waste). The first one (on water) is not due till June 1997 so it is not yet possible to assess how useful they will be, though it is disturbing that the Commission is late in issuing the questionnaires on the basis of which the national reports are to be produced.

One question that will need to be resolved is the role of the EEA in contributing to such reports or undertaking them itself. The Agency has the task of providing the Community and the Member States with 'objective, reliable and comparable information at European level' as the basis for environmental measures, of assessing the results of such measures, and of ensuring that the public is properly informed about the state of the environment. Such tasks could certainly include undertaking the reporting requirements, but since Directives place responsibilities on the Commission, the Directives might have to be amended unless the Agency were to undertake the work on behalf of the Commission. The implementation network IMPEL (see below) could also have a role in contributing to such reports or at least debating them.

In addition to formal reports, independent organisations (NGOs, research institutes, universities, industrial associations) are free to make their own evaluations and comparisons and so contribute to public discussion and awareness of the effects of EC legislation. This can be done effectively only if the basic information is public. The *Manual of Environmental Policy: The EC and Britain* produced by this Institute and updated every six months explains how all EC environmental legislation has been implemented in one Member State and the effects that EC legislation has had in practice. Similar books have been produced for some other countries by our sister institutes. Comparative reports for single or groups of Directives have also been produced.

Administrative and judicial enforcement

The enforcement of Community obligations is primarily a matter for the Member States and their relevant competent authorities. In the words of the UK House of Lords:

> the rigour with which Member States enforce Community law thus in general reflects national enforcement policies, the vigilance and competence of national regulatory agencies and the legal remedies and sanctions available under national law. It is also dependent on the degree to which accurate information about environmental media is collected and handled. Given these variables, it is inevitable that enforcement and its effectiveness will differ across the Community not merely between but within Member States.
>
> *(House of Lords 1991)*

Article 130s(4) of the Treaty confirms that the prime responsibility for implementation rests with the Member States. However the Commission also has a duty under Article 155 to act as 'guardian of the Treaty' and to ensure that measures adopted under the Treaty are applied. This extends not just to transposition but to practical implementation.

One response to the frequently made allegation that the effectiveness of enforcement differs across the Community has been the creation in 1992 of the implementation network known as IMPEL. This is open to representatives of environmental enforcement bodies concerned with industrial installations within the Member States and therefore does not cover all EC environmental legislation (e.g. the birds and habitats Directives, the bathing water Directive, and non-industrial developments under the EIA Directive). The Commission provides one of the joint Chairmen of IMPEL, which is therefore coming to have a semi-official status. An indication of the tasks of IMPEL is given by the four working groups it has established:

- Technical aspects of permitting
- Procedural/legal aspects of permitting
- Compliance monitoring and inspection
- Managing the enforcement process.

In addition there is an ad hoc group on transfrontier shipment of waste.

One of the roles of IMPEL is to provide a forum for professional regulators to exchange information about the details of enforcement methods used in the Member States. They should thereby educate each other and so help disseminate good practice throughout the Community. IMPEL is still young and the results of its work are not yet well known.

There is no Community environmental inspectorate although the idea has been discussed for many years, possibly in the form of an 'inspectorate of inspectorates' (see Parliament's Resolution on implementation of water legislation OJC 94/157 paras 27 and 28, 11.4.88). It will be discussed again when the Regulation establishing the EEA is reviewed. Any new discussion will have to consider not just the powers of such an inspectorate, but also the field it is to cover. Is it, for example, appropriate for a single Community inspectorate to cover both industrial pollution matters and nature conservation?

Prosecution initiated by the competent authorities in the national courts is a last resort after all administrative steps have been taken. If the competent authority fails to initiate proceedings, it is possible in some Member States, depending on national legal traditions, for third parties to bring an action to compel the competent authority to act, or to bring a case for damages under civil law if this can be proved. The rights of access of third parties to the courts varies between Member States, and the Commission is known to be considering a Directive on common rules on who should be entitled to seek judicial review of administrative acts and omissions.

A further possibility open to members of the public, including NGOs, who believe that a Directive is not being correctly implemented is to make a complaint to the Commission. There would be fewer complaints and the Commission would have a lighter task in fulfilling its obligations to act as guardian of the Treaty if all Member States perfectly fulfilled their obligations. The Commission's task falls into two parts: ensuring correct transposition and practical implementation. It examines all the national laws, regulations and administrative provisions sent by the Member States to demonstrate that they have transposed all the provisions of Directives into national law. If necessary, after sending warning letters and a Reasoned Opinion, the Commission can initiate proceedings before the Court of Justice for failure of transposition. The suggestion is sometimes made (e.g. by the House of Lords in the report quoted above) that Article 169 letters, Reasoned Opinions and their responses should be published.

Ensuring practical implementation is a much more difficult task since the Commission does not always have all the relevant information and has no inspectorate able to collect it. The Commission is therefore dependent for its information on those reports required under Directives (see above) and on other sources including complaints made by the public. The EEA should be able to contribute as its tasks include providing 'objective information necessary for framing and implementing sound and effective environmental policies' and drawing up expert reports which the Commission can use 'in its task of ensuring the implementation of Community legislation'.

Meanwhile complaints by the public (individuals, NGOs, industries, local authorities, etc.) are an important source of information for the Commission. It is a practice that has grown, stretching the ability of the Commission to deal with them. The disadvantage of the complaints system is that it may distort the selection of infringements which the Commission deals with, focusing its attention on those complained about to the neglect of possibly more important infringements. Countries with well developed NGOs that have learnt to use the system are subject of more complaints than other countries whose infringements may be more significant. There are, however, two important positive aspects of the complaints system: it exerts pressure on the Member States and competent authorities to take implementation seriously; and it makes a reality of Community legislation for the citizens. Instead of Community legislation being seen to consist of pieces of paper which may or may not have an effect, and therefore may not be worth reading, the interested public has a purpose in studying and understanding EC legislation and knowing that they have a role in making it work. It is indeed as a result of complaints that the Commission has successfully brought several cases before the Court of Justice relating to practical implementation and has thus forced Member States to take action to protect the environment.

One response to the growth in numbers of complaints has been to suggest better means to deal with issues at national level.

Conclusion

This paper is intended to show that implementation has many aspects. The low priority accorded to examining the problems of implementation may indeed result from this fact: it is a more difficult subject to deal with than adopting new legislation, and not as glamorous. To some people it appears backward, instead of forward, looking. Any who think this should reflect that, if this view is correct, then the exciting item of new legislation that is now being discussed will shortly suffer the same fate. The view that the adoption of new legislation is all that matters risks EC policy being nothing more than an impressive array of pieces of paper.

Implementation should be seen much more positively. Implementation is not just a narrow legal matter (transposition), though it includes that, nor is it a narrow technical matter (e.g. monitoring), though it includes that too. Implementation is the link between what is desirable and what is achievable; it provides the essential feedback for the improvement of new policy-making. It also provides the link between Community policy and national policies, and ensures not only that Community policies really penetrate into each Member State, but also that an understanding of national and local policies informs the development of Community policies. EC policy really comes to life only when it is implemented in the Member States and has become inseparably intertwined with national policies.

25 April 1996

Developments since 1996

Some months after the Parliament's Hearing, the Commission issued a Communication on 'Implementing Community Environmental Law' (COM(96)500). This claimed to take 'into account the methodology of the "regulatory chain" which demonstrates in successive stages all the problems related to implementation (legislation, transposition, practical application, enforcement and review)'. The Communication set out to raise awareness of the subject, to reinforce existing actions and to introduce three new areas for action: setting minimum criteria for inspections by Member State bodies; supplementing the procedure of complaints made to the Commission with similar procedures at national level; and enhancing access to justice before national courts. Action has been taken on the first with the help of IMPEL (see Chapter 8) and several years later provisions on access to justice in certain circumstances were included in the 'public participation' Directive 2003/35. The EU is also a party to the 1998 Aarhus Convention (on Access to Information, Public Participation in Decision making and Access to Justice on Environmental Matters) and, although the European Court has recognised the right of access to national courts by the public, the Commission is considering the possibility of further EU legislation on the subject.

The Communication referred to the potential for the EEA to provide information for evaluating the effectiveness of EU legislation.

In 2008 the Commission issued a more comprehensive Communication, COM (2008)773. It identified problems in specific sectors, waste, water, nature, etc. It explained that DG Environment had established internal taskforces in these sectors to tackle problems that exist at a significant scale and across Member States. It reflected on the need for a strategy to prevent breaches in the first place, including guidance documents to help avoid interpretative misunderstandings, and structured dialogue with national authorities. The Commission would help fund judicial training and would exchange information with the European Forum of Judges for the Environment and the Association of European Administrative Judges. Infringement procedures would be used to tackle the most important problems, with criteria set for selecting these. NGOs were recognised as an important source of information. The Commission would continue its dialogue with the European Parliament. The environment was said to account for 10 per cent of all parliamentary questions put to the Commission, and 35 per cent of petitions to the Parliament relate to the environment.

One of the nine priority objectives of the 2013 Seventh Environmental Action Programme is to improve implementation. The Programme noted that the environment was the area of EU law with the most infringement proceedings before the Court and that the costs of failure to implement are high. Efforts to improve matters would focus on certain key areas, sometimes repeating suggestions put forward in the earlier Communications, as follows:

- Knowledge about implementation is to be better disseminated. Partnership implementation agreements between the Commission and individual Member States are foreseen.
- Requirements for inspections will be extended to a wider body of environmental law, and further steps taken to develop inspection support capacity.
- The way in which complaints about implementation of EU are handled at national level will be improved (without saying how). (This repeats one of the three new ideas put forward in the 1996 Communication without saying what experience had been gained in the previous seventeen years.)
- Citizens will have effective access to justice, and non-judicial dispute resolution will be promoted.
- To share good practice, enhanced cooperation at EU level will be promoted between professionals working on environmental protection including Government lawyers, prosecutors, ombudsmen, judges and inspectors. IMPEL is mentioned in this context.

It is clear that the subject of implementation will never go away, despite the steps taken since the subject was first identified as worthy of attention. Some fault lines that create impediments to effective implementation are, for obvious reasons, never mentioned in communications from the Commission. Since the Commission has the power to take Member States before the Court, and since the Court now has the power to issue fines, the Member States are never eager to admit deficiencies. Dialogue between the Commission and the Member States is therefore never going to be easy. Another fault line runs through the Commission. The task of checking transposition rests with the legal unit in DG Environment, and it is that unit that processes complaints from the public about failures of implementation. But 'practical implementation' is primarily a matter for the units in DG Environment responsible for different subject matter areas, whose staff is far too small to do all the necessary work unaided. A third fault line is caused by the usual rivalry that exists between institutions, in this case between the Commission and the EEA. For its own reasons the Commission has been reluctant to see the EEA evaluating legislation. I expressed this rivalry in these tactful terms when I served on the EEA Board, in the foreword to an EEA report on evaluating effectiveness (EEA 2001):

> the report challenges both the European Commission and the European Environment Agency. It is not just a question of what information we need in

> order to evaluate effectiveness, or of how to do it, but also of who does it or at least contributes to doing it.
>
> The Commission has not been systematically evaluating the effectiveness of EU legislation, and the data it has is incomplete to enable it to do so. The Agency has concentrated on collecting information about the state of the environment, and has not been encouraged to collect other relevant data such as on the institutional arrangements within Member States, and what they do to make the legislation work. Yet such data on 'state of action on the environment' is as necessary as that on 'state of the environment'. The Commission has perhaps not encouraged the Agency to collect all the necessary data, because it feels some of it is too close to the making of policy, which is not the Agency's job. It cannot therefore be repeated too often that evaluating effectiveness of existing policy is not the same as making new policy. An evaluation of existing policy is information relevant for the making of new policy, and it is indeed the Agency's task to provide information relevant for the making of policy. Certainly the European Parliament expects the Agency to help it with its own evaluations.

My failure to persuade the EEA to work systematically on implementation was a major disappointment. If, as the debate about implementation continues over the years (see Chapter 15), it is finally decided that more needs to be done on evaluating effectiveness, and if the current moves for the EEA to do more are not realised, then the possibility of establishing a new Executive Agency with that task will gather support. There is no hiding the fact that the task is a large one that will only be accomplished with adequate funding.

Note

1 The reason I repeat this fanciful analogy is because an official from the Council Secretariat sought me out after my presentation to tell me that I had perfectly caught the Council's approach to the adoption of Directives. It rang true to her at least. I have not included the Parliament among the 'parents' because in 1996 it did not have the powers it does today.

References

Bulletin of the EC (1990) 23(6), p 18.

EEA (2001) *Reporting on Environmental Measures: Are We Being Effective?*, Environmental Issue Report No. 25, Luxembourg: Office for Official Publications of the EC/European Environment Agency.

Farmer, A, ed. (2012) *Manual of European Environmental Policy*, London: Routledge. (This Manual has replaced earlier versions. It is available on the IEEP website.)

Haigh, N (1984) *EEC Environmental Policy and Britain – An Essay and a Handbook*, London: Environmental Data Services.

Haigh, N (1997) Effective environment protection: challenges for the implementation of EC law. Background paper for invited experts and participants. In: *House of Lords, Select Committee on the EC, 2nd Report Session 1997–1998, Appendix 5*, London: The Stationery Office.

House of Lords (1991) *The Implementation and Enforcement of Environmental Legislation Select Committee on the EC 9th Report Session 1991–1992*, London: The Stationery Office.

15

RETAINING THE CENTRE STAGE

David Baldock

Introduction

Environmental policy in Europe has not stopped evolving and is most unlikely to do so notwithstanding suggestions that it is now 'mature'. Climate policy, initially a tentative new strand in a panorama of mechanisms to address global concerns, is perhaps the outstanding example. Since the 1990s it has gathered pace, almost outgrowing the confines of mainstream environmental policy to create a domain of its own, extending into energy, transport and economic policy, foreign affairs and elsewhere. In other respects, however, the long period of constructing EU environmental policy from the 1970s onwards reached a turning point in the early years of the new century. New legislation was being added at a slower pace than previously and a measure of maturity had been achieved.

This was unsurprising in the sense that EU policy had been extended to the point that it addressed most of the areas of environmental concern that were considered relevant at the European level and were within the EU's formal legal competence. The environmental *acquis*[1] had become a relatively comprehensive body of legislation through a succession of both systematic endeavours and more unpredictable episodes, outlined earlier in this book. It has moved, in the words of Chapter 1, from obscurity to centre stage. Certain gaps in the scope of policy could be observed, for example in relation to soil protection, which remained the domain of Member States despite Commission efforts to tackle aspects of this question through legislative proposals which aimed to complement the suites of policy on the other principal environmental media of air and water.

Furthermore, new issues continued to arise and the level of implementation of many environmental Directives remained incomplete or insufficient or both. Climate policy had gathered momentum from the 1990s with an interplay between EU policy and global agreements but this was rather exceptional. Indeed, it could be argued that the high-level focus on climate policy took up much of the political oxygen available for environmental issues, leaving less space for progress on other issues such as air quality, biodiversity, resource efficiency, and so on. After decades of fairly intense activity a period of slowing down in the main body of environmental policy was not unexpected.

The explanation for this change of pace lay not only within the dynamics of environmental policy itself but also in the wider affairs of the Union. Ten new countries, predominantly in Central and Eastern Europe but also including Cyprus and Malta, joined the EU. This enlarged the territory and the number of jurisdictions subject to EU environmental law as well as expanding the ranks of decision-makers, the range of national political and physical realities, and thus the political complexion of some of the issues arising. With twenty-eight Member States represented at Council meetings, the conditions in which debates could be held and conclusions arrived at also underwent a change. Given the low incomes, ageing industrial plants, battered infrastructure and ruptured institutions in many of these countries, it was often assumed that the pace of environmental progress in Europe as a whole would need to be slowed down to accommodate the new spectrum of needs in a more diverse Union.

The combination of a reduced momentum within environmental policy and an associated focus on consolidation, occurring alongside the process of EU enlargement, gave rise to a new epoch in the policy, starting around 2003/04. Within five years the global economic recession of 2008 and the political responses to it were to create a further shift in the landscape and led to greater caution in advancing much of the environmental agenda, particularly within the Commission's Secretariat General where it was liable to be seen as a hindrance to the 'growth and jobs' elements of economic recovery.

There has been no decisive break with the past or moment when the corpus of environmental policy has been declared complete. Indeed, the EU has continued to be a source of impetus on many environmental questions, both internally and internationally. EU policy in this field arguably has become the most complete and influential in the developed world. Nonetheless, an era of growth of environmental policy, with the underlying confidence that accompanied it, has passed. A decidedly more challenging era, with a potentially less secure role for EU policy, dawned around a decade or so ago and is now firmly in place.

In this final chapter of the book an attempt is made to look forward to consider the experience of the past decade and to use this as the starting point for some modest speculation about the future of EU environmental policy. In this respect it is a different although complementary enterprise from that of the earlier chapters, as well as being written by a different author.

Enlargement and its aftermath

The environmental *acquis* has not altered significantly in response to enlargement. New Member States were expected to comply with the existing legal requirements, usually from the moment of accession. In some cases special conditions were negotiated, principally transition periods allowing full implementation of a measure to be deferred to a given date.

Chapters on environment policy were amongst the more detailed ones in the respective accession treaties which were negotiated with each new applicant individually. Most governments might have had a good case for negotiating extended transition periods for measures requiring substantial investment, for example in waste water treatment facilities or where they lacked the institutional capacity to implement unfamiliar new measures. However, this did not occur on a large scale and the extent of derogations was relatively modest when measured against the level of adjustment required and the resources available in most countries.

One reason for this was the level of enthusiasm for joining the EU and so accepting the obligations it entailed, but also for achieving a higher quality of life, including higher environmental standards. There were considerable pressures to adopt the *acquis* quickly and for governments to move at the pace required to allow a large group of countries to join at once, although Bulgaria and Romania fell out of the cohort of ten applicants and joined the EU two years later. Staff at the Institute for European Environmental Policy (IEEP) experienced this pressure at first hand in Budapest where we maintained an office for a few years, dedicated to helping the Hungarian Government to adapt to the environmental *acquis* and to transpose relevant EU requirements into national legislation. In this and many other projects in the candidate countries we encountered real enthusiasm for raising environmental standards as part of becoming a modern European society, not least amongst the, mainly young, officials in environment ministries who were often pivotal in overhauling aspects of national law and establishing new procedures and institutional structures.

In some new Member States there was a sense that a large proportion of older factories and industrial facilities were near the end of their lives and, whilst they probably would not be able to comply with EU standards, their projected lifetime might be relatively short. There may also have been more tacit assumptions, for example that the level of prosperity expected to follow EU membership, not to mention the availability of Structural and Cohesion funding which would help to drive that prosperity, would help to offset costs and launch a wave of new productive investment. In any case, demanding environmental legislation was not unfamiliar in the Soviet era; frequently it was not complied with, and for newcomers to the EU it was not always easy to gauge how far the European Commission would pursue infractions against governments that had exhibited good will in signing up to the environmental *acquis* even if they were not in a position to implement it very thoroughly. On top of this, some governments in Central and Eastern Europe were conscious that opponents of enlargement were highly concerned about a shift in industrial investment out of Western Europe into the new Member States, fearing lower environmental and social welfare standards as well as lower labour costs.

With emissions per capita of many industrial pollutants, including greenhouse gases (GHGs), relatively high in some of the new Member States on the EU's eastern flank, and even more so per unit of GDP; and with extensive areas of dereliction, the environmental pressures were evident and visible. On the other hand, the populations of many species of fauna and flora had fared better in the East than in the West and the intensity of agriculture had declined over large areas since the transition from communism in the late 1980s and early 1990s. Many nature reserves had been strictly protected under communist regimes and the heritage of semi-natural habitats was richer than that in more affluent parts of Europe. In many cases the new Member States designated 20 to 30 per cent of their territory as Natura 2000 sites and also could offer substantial scientific expertise in managing them. The challenge, more commonly, was to secure and deploy sufficient resources to manage this heritage sustainably.

Most of the tools required for tackling the environmental challenges in the new Member States were available in EU policy. This is probably one of the reasons why there was less resistance to the environmental *acquis* during the accession negotiations than might have been anticipated given the substantial costs involved in meeting EU standards and the difficulties of doing so within a short period. Some contribution was made to meeting these costs through EU cohesion policy and a range of EU funds directed at closing the gap between richer and poorer economies in Europe. Nonetheless, the issue of compliance costs, already prominent for some measures in southern Europe, was considerably intensified by the 2004 enlargement. At the same time, the existing gap between the standards laid down in a range of environmental legislation and what had been achieved on the ground grew larger and more visible across the EU. Meeting the standards in the urban waste water treatment Directive 91/271 and the water framework Directive 2000/60, for example, in the new Member States would require systematic investment over a prolonged period and an institutional determination previously lacking in most governments.

Not unreasonably, many in the environmental community in the 'older' Member States (referred to as the EU-15 from this point on) feared that enlargement would result in growing resistance to any further increase in environmental ambition or the setting of higher standards. Why should governments with so many pressing socio-economic challenges to meet share the priorities of the more advanced economies in the West?

In practice, however, the new Member States initially were reticent to question new environmental proposals from the Commission any more rigorously than their counterparts in the EU-15. Very often they were supportive of EU measures, even if they might be relatively demanding to implement in practice. Nor did they move rapidly to form a political bloc designed to utilise their votes in the European Council in pursuit of a regional set of interests. Only after a period of years did they become more frequent questioners of the case for higher standards and in this their focus was particularly on measures addressing climate change. Here they shared a number of concerns not only about the cost of action but also the implications of reducing reliance on fossil fuels, which were an integral part of nearly all their economies, and especially so in Poland where coal production as well as combustion had, and continues to have, a special political as well as economic significance.

Politically the most visible aspect of their cooperation was the attention the 'Visegrad Group', consisting of Poland, Hungary and the Czech and Slovak Republics, started to pay to environment and energy issues, often forming a platform to take a joint approach to negotiations on EU policy. The Visegrad Group has become an active faction in the political landscape on the environment, and has exerted influence on the direction of European policy, for example in relation to emission reduction targets, aspects of the Emissions Trading Scheme, mitigation efforts in sectors outside the main industrial emissions, investment aid for infrastructure including transport links, and various other measures. The common stance amongst Central and Eastern European countries, often spearheaded by this group, can make it more difficult to reach a common EU position in certain areas and has led to some special arrangements or concessions to a number of countries, notably Poland, often justified in relation to their particular economic circumstances.

Looking ahead, the need to balance interests in eastern, western and southern parts of the EU, particularly for climate-related and other measures of economic significance, will continue to be an important concern for those designing and indeed implementing policy. The linkage between legislative requirements and budgetary adjustments has become stronger and in this sense the EU reflects some of the dynamism of wider climate negotiations on the world stage, where the distribution of effort between rich and poor and the flow of financial resources between the two is central to unlocking progress.

The EU enlargement also underlined the need to achieve effective transposition and implementation of environmental legislation and the extent to which this could lag behind the pace at which new measures are introduced. The Commission has always striven to demonstrate consistency in the rigour of its approach to Member State implementation. This became even more important with the new range of national environmental and economic realities in which the legislation applied: there were implications for newer and older Member States alike.

Since enlargement coincided with the relative maturity of the *acquis*, noted already, it has helped to shift the strategic focus of environmental policy towards implementation and away from the introduction of new measures. This is signalled in many documents, not least the Seventh Environmental Action Programme 'Living Well, Within the Limits of Our Planet' (Decision 1386/2013) agreed in 2012. This referred to 451 infringement cases related to EU environment legislation in 2009 alone. In practice, improving implementation has proved easier to discuss at a strategic level than to put into practice on the ground. Sensitivities about Commission efforts to strengthen their role in pursuing more rigorous implementation, for example by strengthening monitoring and inspection requirements or taking cases against errant governments to the European Court of Justice, have not diminished.

With twenty-five Member States then participating in decision-making, now twenty-eight with the later accessions of Bulgaria, Romania and Croatia, there is less time available for individual Ministers to contribute to discussion within the Council. In environmental policy, as elsewhere, governance systems have had to adapt, with more business being conducted outside formal meetings of the Council and less opportunity for individual Member States to focus discussion on their specific concerns. The number of officials responsible for environmental policy within the European Commission has not grown commensurate with the enlargement of the Union and the number of governments and other stakeholders involved. The pressure on them has increased and their capacity to familiarise themselves with conditions on the ground or visit a significant number of Member States has declined. To some extent this has been compensated for by much enhanced electronic communication, solid work by the European Environment Agency (EEA) in building an environmental profile of Europe as a whole, and the enhanced deployment of experts, academics and consultants to inform the machinery of government. The exact contours of this change are difficult to map but the dynamics and colour of policy-making in the EU has changed in ways that seem unlikely to be reversed.

The rising climate agenda

Recognition of climate change as a threat to the biology, the habitability and the economy of the planet has grown steadily in Europe over the past fifteen years and, whilst sceptics remain, the consensus in favour of action has been much stronger than in the United States. The EU's engagement with the UN Framework Convention on Climate Change and the negotiations over a global binding agreement on emissions reductions and associated issues have been persistent and initially had some success in contributing to the agreement on the original Climate Convention and the Kyoto Protocol as outlined in Chapter 9. However, the EU's leadership role encountered a major setback at the important Conference of the Parties (COP) to the Convention in Copenhagen in 2009 when progress towards a binding agreement was stalled.

Nonetheless internal EU policies have grown in ambition since the advent of the Kyoto Protocol and have a more concrete goal than applies in many other areas of environmental policy. In 2005 the European Council committed the EU to pursuing policies designed to avoid aggregate global warming above 2°C [COM(2007) 0002]. Progress towards this target, in the EU and elsewhere, falls a long way short of what is required but a succession of EU strategies and plans have been put in place, together with some substantive legal measures, as outlined in Chapter 9. This shows at least a recognition of the scale of action required and provides a strong rationale for working backwards towards implementation of the measures required in the shorter term.

Perhaps the most important of the more recent plans is the Low-Carbon Economy Roadmap for 2050 [COM(2011)0112] outlining the route to attaining an 80–95 per cent reduction in GHG emissions by this date. This Roadmap was supported by all Member States except Poland. Concrete EU measures to advance in this direction have been introduced in a highly politicised context, with considerable disagreement over how fast the EU should move and whether it should commit more or sooner than other OECD countries, notably the USA, and whether it should be prepared to show more ambition in the event of a global deal. At the same time responsibilities have shifted within the Commission, with the creation of a separate Directorate-General for Climate in February 2010. This was taken further in the Juncker Commission with the appointment of Miguel Arias Cañete as the EU Commissioner for both Energy and Climate Action, taking effect from November 2014.

At the heart of climate policy in recent years there have been agreements on emission reduction targets for the EU as a whole and, within this envelope, for individual Member States. An initial set of targets were set for 2020, at a level agreed by most environmental observers to be disappointingly lacking in ambition. In October 2014 a second process of agreeing future EU emission reduction targets was concluded, committing to 'at least' a 40 per cent reduction in overall EU emissions of GHGs by 2030 (EUCO 2014). On this occasion, binding targets for the proportion of energy supplies to be met from renewable sources and for improving energy efficiency were also agreed but were not binding on Member States. It remains to be seen what, if anything, the notion of a target 'binding at EU level' without imposing obligations on Member States actually means.

This is a significant change from the 'Climate Action and Renewable Energy Package', or 20–20–20 agreement, concluded in advance of Copenhagen, which gave considerably less latitude to Member States (CEC 2015a). To meet the 2030 targets it seems likely that governments will be able to devise their own strategies as well as implementing certain policies applying at the EU level, perhaps most notably the European Emissions Trading Scheme (ETS) Directive 2003/87, which was introduced in 2005. For the moment it is the world's largest 'cap-and-trade' system, with many elements that have provided a model others have followed. It applies to larger industrial plants and power stations, and over time is designed to drive down emissions in a cost-effective way on a continental scale. In practice, however, the emissions cap has been set at too high a level for most of the ETS's existence, particularly in relation to lower emissions caused by the economic recession.

Mainly for this reason the ETS has not created a sufficiently high market price for carbon to reflect its environmental externalities, and does not seem to have driven a substantial flow of investment into lower-carbon facilities, other than a switch into gas-fired power stations at the expense of other fossil fuels. Consequently, measures to strengthen the ETS have been initiated, including at the time of writing an agreement on a new Market Stability Reserve (European Parliament News 2015). However, its role as the potentially keystone measure in addressing industrial emissions in Europe has yet to be proven. Further refinement may be required alongside a greater commitment to setting the cap at an appropriate level. Member States have a collective interest in giving the Commission sufficient latitude to operate the scheme in an effective way to squeeze emissions downwards irrespective of their immediate economic interests or commitment to subsidiarity. This is a crucial testing ground for future EU policy.

The Renewable Energy Directive 2009/28 is an example of climate change policy's growing reach. Strictly an energy measure, its principal purpose is to facilitate a timely transition to the decarbonised energy system needed to meet EU climate objectives. Experience with it has been rather different from the ETS in that it obliges Member States to meet a fixed proportion of their national energy supply from renewables by 2020, with different national targets contributing to an overall target of 20 per cent for the entire EU. Many of the national targets demand a substantial level of investment in new renewables, such as wind power and solar, if they are to be met. In response, governments in most of the EU have been obliged to introduce incentive schemes for suppliers in order to generate the necessary level of new capacity and also to start building the necessary supporting infrastructure, including enhanced transmission systems. While it is too soon to establish whether the targets will be met by 2020, it is clear that the Directive has created a leap forward in capacity, transformed the renewables market in Europe and brought down the costs of new renewables. This 'dirigiste' approach has its drawbacks; the costs of some new supplies might be higher than under a more liberal low-carbon energy market, and there has been a disproportionate investment in solid biomass such as wood pellets burned in large power stations. Conversion from coal to biomass is a relatively rapid means of increasing renewables deployment, but raises serious concerns about sustainability and the real level of GHG reductions achieved, which can be lower than they appear under current accounting rules (see Bowyer et al. 2012).

In the ETS and the Renewable Energy Directive the EU has two models of policy both of which have value in reducing emissions. Arguably they are much more complementary than competitive. Both need further development and well-tuned detailed rules to maximise their value. However, they raise different issues with regard to effectiveness and subsidiarity. The EU has found it much easier to make progress on areas of policy which involve the trade of goods across borders, with instruments to improve the emissions performance of cars or the energy efficiency of household equipment proving generally acceptable in principle, however contentious the precise standards proposed. It is not yet clear whether the combination of EU and national policies that will be in place after the current package of rules expires in 2020 will be sufficient to maintain the momentum that has been created since 2008. In any case, climate objectives will need to stretch deeper into energy supply, energy policy and beyond to industrial policy if a robust low-carbon economy is to be created.

Under the policy guidelines set by the European Council for the decade to 2030, Member States will not have binding targets for renewables or for improvements in energy efficiency or for the transport sector, and it is unclear how the EU as a bloc will meet its commitments. Effective governance mechanisms look critical alongside the setting of targets, although the Juncker Commission seems cautious about imposing more than the lightest structures on Member States.

Some of the less discussed challenges include the sectors where GHG emissions are more difficult and often more expensive to reduce substantially, including transport and agriculture. In these areas Member States are opting more for subsidiarity than for common EU-wide rules, with some exceptions, for example in relation to aspects of biofuels and to technical standards for vehicle electrification.

The Juncker-led European Commission, which came into office in October 2014, has seemed content to allow the pendulum to swing away from pan-European measures on climate issues but is more inclined to advance them in pursuit of the prospective 'Energy Union', for which proposals were launched in spring 2015. Climate seems likely to remain one axis along which this tension will continue to be played out, reflecting the parallel negotiations at the global level. With key measures, such as the recent 2030 climate and energy package (CEC 2015a), having to be determined at the highest political level – the European Council – the rules of unanimity that apply in this forum are potentially constraining and could come under increasing strain, as has been demonstrated by Poland's role as a forceful, sceptical outsider in several recent decisions.

Since 2009 the EU has lost some of its previous status as a global leader in international climate negotiations, although it is still an important player. Its offer to cut GHG emissions by 'at least' 40 per cent from the 1990 baseline by 2030 runs considerably ahead of that of most other OECD countries in the lead-up to the 2015 Paris conference, even though more could be achieved without major economic cost, as indicated in the related impact assessment. In this respect the EU is a front-runner, seeking to exert influence on the USA, China, India and other key players more through its own level of commitment than as a pivotal negotiator. The relative predictability of some of its positions is helpful in this context and contrasts with the reluctance to commit exhibited by many other major emitters and the brinkmanship that sometimes follows from this.

It will be a significant achievement if the EU can retain and amplify this front-runner role over the next decade which will be crucial in determining whether climate change can be constrained within safe limits and the low-carbon economy developed fast enough to create a sustainable model for the longer term. Whilst there is understandable disappointment that the EU has lost some of its leadership role, its own capacity to apply effective climate policies amongst 28 countries and to support developing countries on a sufficient scale remains a key test bed of how other nations and blocs will need to address similar challenges. The EU is a laboratory for developing the necessary policies and supporting measures, such as trading and transfer mechanisms between countries, so as to compensate for differential levels of effort and varying economic capacity. Environment policy and its climate penumbra are clearly centre stage in this respect.

Wider environment policy

The dynamism of climate policy has been less apparent in other aspects of EU environment policy, particularly since the onset of the recession in 2008. Major new proposals are now infrequent and efforts to agree a framework Directive relating to the quality of soils have lapsed. The overall emphasis in policy has shifted towards the following:

- Filling gaps in the existing *acquis*, many of which are relatively technical.
- Improved implementation of existing measures (see Chapter 14).
- Consolidation, and tidying up of existing legal measures, with an eye to simplification and reducing administrative burdens.
- Review and appraisal of the established *acquis* examining its fitness for purpose and considering the case for either revision or for additional measures.
- Pursuing the 'green economy', promoting the larger-scale development and deployment of clean technologies, the growth of green jobs and a stronger environmental focus in key economic sectors including fisheries and agriculture (EEA 2013). Low-carbon goals have been the principal but not the exclusive environmental priorities in this development. The language of the 'green economy' has become more prominent than that of environmental 'integration' in previous years, although a substantial reform of the Common Fisheries Policy was achieved in 2013 (Regulation 1380/2013) and this agenda remains critical. (Without questioning the need to accelerate environmental investment, it is worth remarking that there are clear dangers to treating Green Growth as a central justification for environmental policy. The primary goal of this body of policy is to protect the environment. There is a risk of implying that measures vital for the long-term wellbeing of our natural resources are not worth pursuing unless there is a fairly short-term jobs and growth dividend.)

Some of these themes are signalled in the Seventh and most recent Environmental Action Programme (EAP) (Decision 1386/2013) which offers an overarching framework for EU environment policy from, effectively, 2013 to 2020. Although the 6th EAP expired in July 2012, DG Environment was initially unenthusiastic about proposing a successor and did so only after coming under considerable pressure from the Council, notably under the Danish Presidency, from the European Parliament and from environmental stakeholders, including the NGOs. IEEP organised a number of workshops on the topic and prepared papers which were exchanged with the Commission.

One reason for the reticence was concern about the potential overlap between a new programme and other EU strategic documents or framing measures which have grown in number since the turn of the century. One of these is the Roadmap to a low carbon economy in 2050, referred to earlier. Others include the EU Biodiversity Strategy to 2020 [COM(2011)0244] and one of the key initiatives of Janez Potočnik as Environment Commissioner, the wide-ranging Roadmap to a resource-efficient Europe [COM(2011)0571]. These share a timeframe to 2020 and thus are aligned with the EU's budgetary cycle and the target dates for GHG emission reductions [COM(2012)710].

A further framing measure outside environmental policy and intended to steer the EU's response to the recession beginning in 2008 is the 'Europe 2020 Strategy'

[COM(2010)2020]. This aims for 'Smart, Sustainable and Inclusive Growth' over the medium term and was launched by the Barosso Commission as the keystone document to guide much of policy development in the Union. In political terms it eclipsed the established Sustainable Development Strategy [COM(2009)0400]. It refers explicitly to two sets of environmental priorities, including the package of EU climate-related targets for 2020 mentioned above and in Chapter 9 (the 20–20–20 package) and the more broadly framed resource efficiency goals, notably to completely decouple economic growth from the use of energy and natural resources.

The matrix of strategies that have been put in place reflects a greater recognition of the long-term environmental challenges for Europe, but it also coincides with a further retreat from binding regulation as the preferred mode of intervention on the environment in favour of alternatives. These include cooperative, self-regulatory approaches, more use of information and support, and various kinds of economic measures including limited use of taxes and charges (Volkery et al. 2012).

The 7th EAP is a shorter and less ambitious document than the notable 5th EAP (discussed in Chapter 1), which was prepared at a time when environmental policy was still in an expansionary period. It contains relatively few concrete proposals, and some of those which it does anticipate, such as setting targets for land use and soil, have yet to appear and seem unlikely to do so in the near future. It does, however, offer a coherent and readable account of the environmental challenges facing the EU and the world more generally, noting for example that there is likely to be a global shortfall of 40 per cent in water by 2030 unless there is significant progress in improving resource efficiency. It also sets out some directions of travel, arranged as nine priority objectives beneath which a series of more specific topics are addressed and a limited number of concrete actions proposed. The priority objectives include to:

- protect, conserve and enhance the EU's natural capital;
- turn the EU into a resource-efficient, green and competitive low-carbon economy;
- secure investment for environment and climate policy and get the prices right;
- enhance the sustainability of EU cities.

In some areas of policy there is the clear intention to progress towards longer-term goals by introducing a number of new EU policies. Perhaps the most prominent example of this relates to resource efficiency (see Chapter 6). The 7th EAP proposes that 'targets for reducing the overall lifecycle environmental impact of consumption will be set, in particular in the food, housing and mobility sectors' (p 184). There will be 'measures to increase the supply of environmentally sustainable products and stimulate a significant shift in consumer demand for these products' (p 183). Furthermore, 'Barriers facing recycling activities in the EU internal market should be removed and existing prevention, re-use, recycling, recovery and landfill diversion targets renewed so as to move towards a "circular" economy, with a cascading use of resources and residual waste close to zero' (p 184).

This theme was taken forward by the Commission with a package of proposals on the circular economy in July 2014, including legislative proposals to review recycling and other waste-related targets in the EU waste framework Directive 2008/98, the landfill Directive 1999/31 and the packaging and packaging waste Directive 1994/62. Amongst the proposals was an increase in the target for recycling and preparing for reuse of municipal waste to 70 per cent by 2030. This package would represent a step forward in terms of ambition but has encountered opposition, both amongst certain Member States and within the Commission, as discussed briefly below.

In other areas of environmental policy there is much more emphasis within the 7th EAP and subsequent policy statements on achieving results through better implementation of existing legislation rather than through launching new measures. For example, the EU Biodiversity Strategy [COM(2011)0244] sets the target of no net loss of biodiversity in the EU by 2020, which would require a significant change from present trends given the continued decline of biodiversity and the degradation of many ecosystems. The 7th EAP notes that the Biodiversity Strategy includes (and indeed substantially relies on) measures to improve the implementation of the birds and habitats Directives and states that reaching the headline target 'will require the full implementation of all existing legislation aimed at protecting natural capital' (Decision 1386/2013, p 179).

Improving the implementation of the EU environment *acquis* at Member State level will be given top priority in the coming years. Various means of achieving this are proposed, including improved knowledge of the topic and wider communication of the issues; more tailored, bilateral discussion with individual Member States; extended requirements on inspections and surveillance to the wider body of EU environment law; improved handling of complaints at the national level; better access to justice in environmental matters for EU citizens; and the promotion of non-judicial conflict resolution. In parallel the Commission will 'continue to do its part to ensure that legislation is fit for purpose and reflects the latest science' (Decision 1386/2013, p 190). Regulations may be proposed more often in place of Directives where legal obligations are sufficiently clear, potentially leading to fewer inconsistencies in implementation.

Some of these ideas have been taken up subsequently but not with the boldness or rigour that the 7th EAP implies. There has not been a clamour for more effective implementation either from the European Parliament or from the Council, so proposals to improve monitoring and reporting of implementation within the Member States have been advanced by the Commission with some caution.

The Juncker Commission, coming into office from late 2014, has departed substantially from the tone and aspirations of the 7th EAP, giving priority to a series of 'Political Guidelines' presented to the European Parliament by Juncker prior to his appointment as President of the Commission. This was particularly clear in his 'Mission Letter' (CEC 2014a) sent to Karmenu Vella, appointed as the Commissioner for Environment, Maritime Affairs and Fisheries, a new portfolio covering two separate Commission DGs (which are not merged but overseen by a single Commissioner supported by his cabinet of advisers). This letter, sent on 10 November, argued that

> The EU has a well-developed environment policy with a rather complete and mature legal framework. Protecting the environment and maintaining our competiveness go hand in hand, and environment policy also plays a key role in creating jobs and stimulating investment. This is why the 'Green Growth' approach to environmental policy should be further developed …

Vella was asked to focus on five priorities, the first of which was to overhaul the existing environmental legislative framework 'to make it fit for purpose'. Additionally, he was to carry out an in-depth evaluation of the birds and habitats Directives and assess the potential for 'merging them into a more modern piece of legislation'. This instruction caused consternation in the environmental community, and environment ministries in the Member States appeared surprised. It was particularly challenging to the previous priorities in that it appeared to require a highly contentious conclusion to the Commission's 'Fitness Check' of the two principal nature Directives, a process already under way before Juncker took office (CEC 2014b).

Since late 2014 the priorities of the new Commission have remained jobs and growth, including the launch of a form of economic recovery programme. Whether this amounts to a decisive change of direction away from a broadly supportive view of environmental policy remains to be seen. The immediate rearrangement of responsibilities inside the Commission was difficult to interpret as other than a downgrading of the environment. The Directorates-General for Environment and Marine (covering marine affairs and fisheries) now share a single Commissioner, as do DG CLIMA and DG Energy. Both Commissioners need the explicit support of certain higher-level Commission Vice Presidents with broader, more strategic portfolios in order to progress new initiatives. Initially the portfolios of these Vice Presidents made no reference to the green element in the Green Growth agenda. However, following protests from a number of senior MEPs, the already large span of responsibilities falling within the portfolio of Vice President Frans Timmermans was further expanded to include 'all policies linked to sustainable development' (EESC 2014). Timmermans is the Senior Vice President and a powerful figure in the new Commission, with lead responsibility for pursuing 'Better Regulation' within EU policies and laws, including the environmental *acquis*.

The directions set out in the 7th EAP have not been prominent in the opening months of the new Commission, and priority seems to be given to a document drawn up by Juncker entirely outside the Community process of policy-making. The 'Political Guidelines for the next European Commission' were prepared by Juncker and presented to the European Parliament on 18 July 2014. They contain a ten-point Agenda for Jobs, Growth, Fairness and Democratic change, and do not refer to the environment at all. They were followed by the Commission's draft first year's work programme which proposed the withdrawal of certain environmental proposals including the 'circular economy' package referred to previously (EESC 2014). At this stage there seemed little doubt that the new Commission was embarking on a confrontational pathway with the environmental status quo, looking for a lighter touch and active withdrawal of certain measures.

The political legitimacy of this approach was far from clear and within weeks there was criticism from the European Parliament, the Environment Council and, most vociferously, from some leading environmental NGOs. The German Environment Minister had noted already that merging the birds and habitats Directives was not a proposal viewed with favour, and the Environment Council made it clear that they did not support the withdrawal of the circular economy (Birdlife 2014). The Commission was obliged to retreat, constructing a fudge whereby the annual work programme was accepted largely as it stood but a commitment was made to introduce a new circular economy policy package with a stronger economic dimension during 2015. A public consultation was launched in May, at the time of going to press for this book, and new proposals were expected later in the year.

It is too early in the life of the new Commission to judge whether it is looking for a radical change in EU environment policy in the direction of less binding regulation and more freedom for Member States to pursue their preferred politics within a lighter EU framework. Several of the political signals of the past few months suggest that this may be the case but that it is likely to encounter severe resistance. This is likely to be not just from environment Ministers, many MEPs and NGOs, but also from those industries that depend on binding EU regulations to create either the market conditions for 'green' products, such as renewable energy, or some consistency and certainty within the single market of 28 countries.

Given this level of support for EU environmental policy and the catalogue of challenges to be addressed, including the continued decline of biodiversity, global warming and increased pressure on water resources, it seems unlikely that European society in the wider sense will opt for a substantially lower level of environmental ambition. An intense debate about Europe's willingness to be a front-runner, where necessary moving forward ahead of its partners in other countries, is to be welcomed. The evidence does not support the assumption that more ambitious levels of environmental regulation destroy economic value (OECD 2014; Dechezleprêtre and Sato 2014).

At the same time there will be a legitimate debate about the role of the EU in setting out targets and ambitions on one hand, and precise mechanisms for achieving them on the other. Since 2004/05 the evolution of policy has been towards a larger number of strategies and 'roadmaps', supported by a wider range of policies to advance them. Binding regulations remain an important tool, for example in relation to vehicle emissions, but they are being embedded in a structure that is frequently less centralised. More use has been made of framework legislation such as the Waste Framework Directives, of committee procedures ('comitology'), and of economic instruments, such as the ETS. Self-regulation measures and those relying on information and coordination have a larger role. So does the 'semester' process, involving a dialogue between the Member States and the Commission within a framework established by the Europe 2020 process, with largely economic goals. Environmental policies increasingly have needed to demonstrate legitimacy and relevance in short-run economic terms, and this seems unlikely to be reversed in the foreseeable future.

The emphasis on reviewing EU legislation has also become more prominent, championed by the Secretariat General within the Commission for several years and pushed with renewed rigour by the new Commission. A major package of detailed proposals and procedures in pursuit of Better Regulation was launched by the Commission on 19 May 2015 (CEC 2015b). Better Regulation was described as designing EU policies and laws so that they achieve their objectives at minimum cost and are implemented and reviewed in an open manner, based on the available evidence, with stakeholder involvement. This builds on an agenda stretching back to the Commission's 2001 White Paper on European Governance [COM(2001)0428] and the more recent introduction of 'Fitness Checks', which are relatively comprehensive evaluations of policy within a sector. When four policy areas were selected for pilot Fitness Checks in 2010, the environment was one of them, on the topic of freshwater policy. A much higher-profile review of the birds and habitats Directives began in 2014.

In conclusion

The EU has established a body of environmental law and policy that applies not only in the twenty-eight Member States, but in large part within neighbouring countries in the EEA and the European Free Trade Association. Because of its scope and the considerable number of product standards that it contains, some of this influence extends to trading partner countries as well. Gaps remain: some will be addressed and new issues will arise. However, it seems less likely that EU policy will extend much further into topics that are still the preserve of Member States, such as land-use planning.

On the other hand the process of 'greening' key areas of EU policy that have a major environmental impact, particularly the Common Agricultural Policy, is far from complete. The Common Fisheries Policy was reframed significantly in 2013; but management of the marine environment as a whole, including polluting substances, marine litter, conservation values, marine nature reserves and industrial development as well as commercial fish stocks, remains a major challenge lagging behind what has been achieved in the terrestrial environment. Energy policy increasingly is influenced by climate objectives but this process needs to be taken much further if the level of decarbonisation required by 2050 is to be obtained. A more circular, greener economy will not be achieved without a new generation of policies which will need to include regulation, economic instruments and an imaginative range of new measures.

Environmental policy in a wider sense is certainly far from complete and will need to be taken to a further stage if we are to address the challenges outlined by all the major economic institutions, such as the World Bank and the OECD, as well as environmental agencies such as the UN Environment Programme and the EEA. The current policy is not above scrutiny and improvement, and a measured process of review is appropriate. This should not obscure, however, the scale of the challenges ahead.

Note

1 This is the body of European Union environmental law, legal acts and associated court decisions.

References

Birdlife (2014) Letter from Barbara Hendricks, Federal Minister for the Environment, nature Conservation, Building and Nuclear Safety, to Jean-Claude Juncker, President-elect of the European Commission. www.birdlife.org/sites/default/files/attachments/20140929_letterGEtoJuncker.pdf (Accessed 27. 05. 15).

Bowyer, C, Baldock, D, Kretschmer, B and Polakova, J (2012) *The GHG Emissions Intensity of Bioenergy: Does Bioenergy Have a Role to Play in Reducing GHG Emissions of Europe's Economy?*London: Institute for European Environmental Policy.

CEC (2014a) Mission letter from Jean-Claude Juncker, President of the European Commission, to Karmenu Vella, Commissioner for Environment, Maritime Affairs and Fisheries. http://ec.europa.eu/commission/sites/cwt/files/commissioner_mission_letters/vella_en.pdf (Accessed 27. 05. 15).

CEC (2014b) Fitness Check Mandate for Nature Legislation. http://ec.europa.eu/environment/nature/legislation/fitness_check/docs/Mandate%20for%20Nature%20Legislation.pdf (Accessed 28. 05. 15).

CEC (2015a) The 2020 climate and energy package. http://ec.europa.eu/clima/policies/package/index_en.htm (Accessed 27. 05. 15).

CEC (2015b) Better Regulation. http://ec.europa.eu/smart-regulation/index_en.htm (Accessed 27. 05. 15).

Dechezleprêtre, A and Sato, M (2014) *The Impacts of Environmental Regulations on Competitiveness*, Policy Brief, London/Seoul: LSE Grantham Research Institute on Climate Change and the Environment/Global Green Growth Institute.

EEA (2013) *Towards a Green Economy in Europe – EU Environmental Policy Targets and Objectives 2010–2050*, Report No. 8/2013, Copenhagen: European Environment Agency.

EESC (2014) Jean-Claude Juncker Political Guidelines, Opening Statements: A New Start for Europe: My Agenda for Jobs, Growth, Fairness and Democratic Change and Setting Europe in Motion. www.eesc.europa.eu/resources/docs/jean-claude-juncker---political-guidelines.pdf (Accessed 27. 05. 15).

EUCO (2014) *European Council Conclusions of 23/24 October 2014*, EUCO 169/14. Brussels: European Council.

European Parliament News (2015) Parliament adopts CO_2 market stability reserve. www.europarl.europa.eu/news/en/news-room/content/20150703IPR73913/html/Parliament-adopts-CO2-market-stability-reserve. (Accessed 27. 05. 15).

OECD (2014) *Green Growth – Environmental Policies and Productivity Can Work Together*, OECD Policy Brief, Paris: Organisation for Economic Co-operation and Development.

Volkery, A, Withana, S, Fedrigo, D and Baldock, D (2012) *Towards a 7th Environment Action Programme: Priorities and Action Needs.* London/Brussels: Institute for European Environmental Policy.

APPENDIX

The role of IEEP and the CF/IEEP project in developing integrated pollution control

The 'shift in focus' described in Chapter 8 that led to the integrated pollution prevention and control (IPPC) Directive 96/61 was not easily accepted in the EU despite moves in that direction in some Member States. The inertia of deeply embedded traditions had to be overcome and a way through the difficulties found. How it happened is a tale worth telling, not least because it has lessons for the making of EU policy generally. According to one account the Directive 'originally stemmed from a British attempt to upload a "good" idea to the EU', and the role of the Institute for European Environmental Policy (IEEP) was to help the Department of the Environment (DoE) 'to complete the export process' (Jordan 2002, pp 152, 161). I can well believe that some British officials believed this, and indeed the more people who claim paternity the better, as it affirms their commitment. It does not follow that they will pass a stringent paternity test. It pleased me, for example, when I heard some French commentators claim that the Napoleonic Law of 1810 on *installations classées* already embodied integration since it made no distinction between the environmental media affected by an installation.[1]

The form of EU-wide integrated permitting of industrial plants that was floated at the concluding conference of the Conservation Foundation (CF)/IEEP project in November 1988 differs in one very important respect from what is in the British Act of 1990, and why this difference was so important for success in the EU is discussed below. The form of integration proposed by CF/IEEP not only survived the process of negotiating the Directive, but also survived the review of the working of the Directive some years later. The British 'good' idea as set out in the 1990 Act would never have been accepted by some Member States since it involved a single authority granting a single permit or authorisation which would have run counter to deeply embedded practices in more than one Member State, for reasons explained below. In any event the British were in no position to begin 'uploading' to the EU any ideas on integrated permitting until they had taken the key decision that discharges to water of dangerous chemicals should be subject to strict technology-based controls in the same way as discharges to air. That decision was not taken until November 1987, one year after the CF/IEEP project started. A press article (ENDS Report 154/November 1987) makes clear that the decision was taken under international pressure:

> After a 12-year war of attrition with its European neighbours about the manner in which discharges of dangerous substances should be controlled, the UK is to shift its ground and apply strict technology-based controls to effluents containing the most hazardous chemicals. The Government's decision was announced to both Houses of Parliament on 18 November in the build up to the ministerial conference on the North Sea.[2]

An alternative view is that the CF/IEEP project came up with a 'good' idea for an 'integrated permitting' Directive which was put forward at a conference in Brussels in November 1988, which the relevant Commission officials accepted one year later after they had asked IEEP to conduct a further study (IEEP 1989), and which the UK Government then put its weight behind after the UK Act of 1990 was adopted. The CF/IEEP 'good' idea did not require a single permit granted by a single authority, but allowed different authorities to coordinate their permits to achieve the same result. Had the UK been passive it is likely that a Directive would have been agreed anyway. It was only after the 1990 Act that the UK played an important role. Jordan's account confirms that the UK only became engaged in 1990: 'when the DoE began to pursue a more engaged European strategy in the early 1990s … [it] immediately seized on IPC as just the sort of national innovation that could usefully be uploaded to the EU' (Jordan 2002, p 161). However, it did not succeed in 'uploading' intact its own version – if indeed that had ever been its intention – and had to accept a number of other important changes. As well as the option to have more than one authority issuing permits, the use of energy and resources also had to be taken into account (Emmott and Haigh 1996).

For those who believe that the only actors who make EU policy are the EU institutions and the Member States, then the UK will be regarded as originating the IPPC Directive. It certainly deserves the credit for being the first Member State to use the phrase 'integrated pollution control' in legislation and to require it for industrial plants. But it remains important for an understanding of the EU, and for its reputation as an open form of governance, to insist that civil society also plays a role. There are many examples of environmental legislation being triggered by public campaigns,[3] and examples of non-governmental bodies proposing solutions as well as drawing attention to problems. The IPPC example shows that non-governmental bodies can put forward ideas designed to anticipate difficulties by showing some understanding of differing national practices.

The use of the word 'uploading' also gives a misleading view of EU policy-making. Certainly one Member State often provides the impulse needed for new EU legislation and puts forward detailed suggestions. These are more likely to be successful if they recognise in advance how the other Member States are likely to react and why. We have already seen how the first draft of the 'large combustion plants' Directive 88/609 was modelled on German ideas, but was then significantly changed and improved under pressure from other Member States before being adopted (Chapter 4). The process of negotiation enables the needs of all Member States to be accommodated if at all possible. The outcome is rarely, if ever, the lowest common denominator and, in the case of the large combustion plant Directive, the result was a European solution to the acid rain problem previously unknown in the legislation of any Member State. The process can be creative, and it can be argued that this was the case with the IPPC Directive.

The CF/IEEP project developed as follows. Terry Davies, Executive Vice-President of CF, had remained deeply committed to integrated pollution control ever since, as a staff member of the Council on Environmental Quality, he had drafted the passage quoted in Chapter 8 for President Nixon's 1970 statement to the US Congress. He had also drafted the 1976 US Toxic Substances Control Act which was to influence EU legislation on chemicals (Chapter 7). Terry Davies and I first met at an OECD workshop on cross-media pollution held by its Chemical Group in Paris in September 1985, at which he was quite obviously the best-informed participant. I had been invited to the workshop by Jim Brydon, the head of the Group, to repeat a presentation on coordinating standards for chemicals in the environment that he had heard me give at a conference a few months earlier.[4] After the workshop Davies suggested that CF and IEEP should work together on the subject.[5] We discussed the need to review the existing administrative structures and practices for pollution control (air, water, waste, chemicals) in a selection of European countries to discover to what extent they hindered or facilitated a cross-media approach. The importance of administrative structures had hardly been recognised at the workshop and, as our project was to show, it was an essential factor for success.

My own interest in the subject had been sharpened when looking for a way to resolve the conflict between the UK and other Member States that occurred when the 'dangerous substances in water' Directive 76/464 was being negotiated. The UK had insisted that authorisations (or permits) for discharges to water should be based on the desired quality of the receiving waters and not on the best available technology (see Chapter 5). A proposed resolution of this conflict (Haigh 1984, Appendix 4) was taken up by a parliamentary committee, and when developing my thoughts I had spoken to many officials and industrialists to learn why water and air were treated so differently. A discussion with an experienced water official has stuck in my memory ever since. When he said 'we have always done it like this' and I had pointed out that his air colleagues did it differently, he replied 'oh, but Nigel, we do not talk to them. They are on a different floor.' The reply was intended to be simultaneously apologetic and humorous, but it contained a truth about bureaucracies that is not confined to any one country: officials in one administrative department do not like interference from other departments, and in turn prefer not to interfere themselves. By keeping to themselves they do not learn from each other. Cooperation, let alone 'integration', does not happen easily unless formally required.

The project was managed by Frances Irwin (CF) and myself (IEEP) with Terry Davies (CF) and Konrad von Moltke (CF) closely involved.[6] We started in November 1986 once funding was secured.[7] Our advisory group meeting in Washington in April 1987 selected a number of case studies from a longer list. Preliminary results were reviewed by the advisory group in London one year later and the finished studies were presented at a concluding conference in Brussels in November 1988. One part of the ensuing book (Haigh and Irwin 1990) includes the case studies covering integrated permitting in the Netherlands, Germany, UK and Sweden and one covering policy planning in the Netherlands. Another part describes experiences in the USA and Canada. The British case study, by Susan Owens, titled 'The unified pollution inspectorate and best practicable environmental option', provides an account of the convoluted steps by which the call for greater integration that the Royal Commission on Environmental Pollution had put forward in 1976 began to be put into effect (Owens 1990). Anyone who, with the benefit of hindsight, thinks that the problem is self-evident and the solution follows easily need only read that case study. In the first chapter Frances Irwin drew on the case studies and discussions at the seminar to explain in more depth the need to shift the focus of decisions to 'substance', 'source' and 'region'. The second chapter set out options for the EU, OECD, USA and developing countries, drawing on the case studies and the discussion at the concluding seminar. The project was certainly not confined to the EU.

EU Commission officials attended both our advisory committee meetings. Stanley Johnson, who had already produced the first draft of the Fourth Environmental Action Programme, which included a passage on integration, attended the first. Paola Testori-Coggi of the chemicals unit of DG Environment attended the second and secured funding for the concluding conference in Brussels in November 1988. She asked IEEP to study other Member States and to make specific recommendations for EU action. The resulting report was delivered in November 1989 and repeated the proposal for an integrated permitting Directive as the most promising first step (IEEP 1989). That conclusion was immediately accepted by DG Environment, which could see that several Member States were developing some form of integration and there was a risk that the EU would be left behind. Unfortunately Commissioner Ripa di Meana decided that legislation on environmental auditing should have priority so drafting was delayed (this resulted in the EMAS [Eco-Management and Audit Scheme] Regulation 1836/93). After our conference in November 1988, Jim Brydon asked Frances Irwin and myself to act as consultants to OECD to help him conclude the OECD project on cross-media pollution that he had started in 1985. As a result Frances Irwin and I produced a report and the first draft of the OECD Council Act so that CF's ideas on shifting the focus to 'source', 'substance' and 'region' became official guidance. We also assisted at the OECD meeting in June 1990, chaired by Gareth Bendon of the UK Department of the Environment (DoE), at which the OECD Council Act was finalised (OECD 1991). Gareth Bendon was the official responsible for the drafting of the UK 1990 Act. He had also spoken at our Brussels seminar in November 1988 to explain how British ideas were then developing. In 1991 a British official was seconded to the Commission to begin drafting a Directive and the DoE asked IEEP to write several papers to sharpen their thinking, some about developments in other Member States. Thus all the most motivated parties – OECD, EU Commission, UK, CF/IEEP – were in contact with each other more or less continuously after the 1988 seminar. The Directive was formally proposed in September 1993 and adopted three years later.

It is worth observing that it was officials responsible for chemicals policy in both the OECD and the EU who took the most interest in promoting integrated pollution control. Their colleagues responsible for air and water were more cautious – indeed some water officials were antagonistic. The focus on the 'substance' approach was an opportunity to enhance the relevance of chemicals policy. To the air and water people, integration could be seen as a threat to established ways.

At the final session of our November 1988 conference Frances Irwin and I had drawn conclusions, my role being confined to possible EU action. I started by drawing lessons from the case studies. Some form of integrated permitting was already being practised or proposed in the four Member States that we had studied. In Sweden a single permit covering discharges to all media was already issued at national level, and in the UK it was proposed. In Germany (or rather its western part), with its federal structure under which the implementation of all laws was carried out by the Länder, it would be impossible for a single national body to issue permits. In the Netherlands there were other reasons why that would be difficult. Discharges to the Rhine and the sea were authorised by the Ministry of Transport and Infrastructure and not by the Ministry of the Environment or the Provinces who between them authorised most other emissions to water and air. The reason for this is that the infrastructure Ministry was responsible for sea defences and for dredging the Rhine, and wanted control over the dredged materials that had to be deposited somewhere. To prise this role from such a powerful Ministry – in whose hands the very survival of the Netherlands resides – was not worth even contemplating.

I continued by analysing the EU's fourth Action Programme on the Environment, which included a complete section on integrated pollution control with subsections on the 'substance' and 'source' approaches.[8] The Programme said the Commission would reflect on whether to set controls for particular industrial sectors and that a 'global multi-media, multi-pollution approach' would unavoidably have institutional implications in Member States. It went on to say that 'a powerful unified control authority able to arbitrate as between different environmental sectors to secure optimum solutions would seem to be an unavoidable corollary in such an approach'. I speculated whether this implied a Directive, and suggested three possible forms such an 'integrated permitting' Directive might take, as follows.

- One placing a duty on Member States to establish a 'powerful unified control authority' at national level. This possibility could be dismissed immediately as it conflicted with the federal or regional structures of several Member States.
- One placing a duty on Member States to establish such an authority at whatever sub-national level they deemed appropriate. This would be more within the realms of possibility but would conflict, for example, with the arrangements in the Netherlands.
- A third, more modest possibility would place a duty on whoever grants any authorisation relating to discharges to any of the three environmental media to consider effects on the other media, even if authorisations are granted by more than one authority. This would involve coordination with the objective of minimising the impact on the environment as a whole.

I went on to say that the proposed Directive would also have to remove any blockages to integration in existing Directives dealing with the individual media such as the 'emissions to air' Directive 84/360 and the 'dangerous substances in water' Directive 76/464. Some amendment would anyway have to be made so that these Directives did not prevent Member States adopting integrated permitting.

I was hesitant about the third, more modest, proposal and discussed it with Irwin, Davies and von Moltke before making it. Was such an apparently flimsy proposal strong enough to support a Directive intended to embody a significant shift in EU pollution policy? We could think of none better and so the idea was floated. After conducting more case studies covering other Member States (France, Italy, Denmark and Belgium) and clarifying the others that we had already conducted, one year later IEEP reiterated the proposal in the report to DG Environment (IEEP 1989), which then accepted it. The idea was not welcomed by all.

Gareth Bendon of the UK had written generously after the 1988 conference to say it had performed a 'very useful function in facilitating a meeting of minds on a subject which is poorly defined and only in the first stages of development' but that he was depressed 'about the evident obstacles to the integrated approach and the fundamental nature of some of them which the seminar revealed'. He thought my analysis brought these out well but then said 'your proposition for requiring single medium pollution control authorities to take a cross-media approach is impracticable' and went on to say why: the authorities would not have the expertise or resources to make such judgements. Bendon's comments confirm that the CF/IEEP project had put forward the idea of an 'integrated permitting' Directive before the UK was even thinking of a Directive. Bendon's objection had real substance, but I like to think that the subsequent establishment of the IPPC Bureau in Seville, for which CF and IEEP can claim no credit, will have allayed those reservations. Despite such reservations – and there would have been others – our proposal survived the process of drafting and negotiation to surprise us all by becoming the cornerstone of an important part of EU environmental policy. Terry Davies's aspiration for the US Environmental Protection Agency expressed in 1970 thus took root across the Atlantic a quarter of a century later.

Notes

1 The 1810 law required certain installations to be authorised. They were to be classified under three categories: those which ought to be distant from private dwellings; those which ought not to adjoin dwellings unless conducted so as to cause no nuisance; and those which might remain without inconvenience near dwellings. The 1810 law – much amended – has remained the foundation of the regulation of industrial plants in France, and indeed of several European countries conquered by Napoleon.
2 The ENDS article continues: 'Two individuals have contributed most to the UK's change of heart. Nigel Haigh of IEEP has made the case at length for a marriage of the best of the UK's and continental systems, and his argument was taken up by a House of Lords' Select Committee in 1985 (ENDS Report 127, pp 11–12). And it was William Waldegrave, then Under Secretary of State for the Environment, who opened the door to a shift in policy (ENDS Report 126, pp 17–18) and ensured that it gathered momentum in the ensuing months.'
3 Directive 83/129 banning the importation of the skins of seal pups was adopted following a petition to the European Parliament accompanied by 3 million signatures including that of Brigitte Bardot. The 'birds' Directive 79/409 was also the result of public agitation against the annual slaughter of migratory birds in some countries.
4 The conference was on regulating existing chemicals and was organised by the European Environmental Bureau (EEB) in June 1985.
5 CF and IEEP had already collaborated in the late 1970s on a comparison of US and European approaches to controlling chemicals. It was probably then that Konrad von Moltke learnt from CF about the need to integrate pollution control. Certainly it was a talking point when I joined IEEP in 1980.
6 Konrad von Moltke had by then left IEEP to live in the USA, where he was associated with CF. He conducted the German case study.
7 The German Marshall Fund of the USA made the biggest contribution. The European Cultural Foundation funded some of the case studies. The EU Commission funded the concluding seminar in Brussels, conveniently fulfilling the promise in the Fourth Action Programme that the Commission would reflect on integrated pollution control.
8 The Fourth Action Programme was formally proposed in October 1986 and adopted in October 1987. The first draft was written by Stanley Johnson, a British official in DG Environment who would certainly have been familiar with the British Royal Commission's report of 1976. He will also have known of the ideas being discussed within OECD including CF's 'shift in focus' to 'substance', 'source' and 'region'.

References

Emmott, N and Haigh, N (1996) Integrated pollution prevention and control: UK and EC approaches and possible next steps, *Journal of Environmental Law*, 8(2), pp 301–311.

ENDS Report (1987) *Major UK Policy Shift on Discharges of Dangerous Substances*, ENDS Report 154, November 1987.

Haigh, N (1984) *EEC Environmental Policy and Britain – An Essay and a Handbook*, London: ENDS.

Haigh, N and Irwin, F (1990) *Integrated Pollution Control in Europe and North America*, Washington, DC: Conservation Foundation.

IEEP (1989) *Possibility for the Development of a Community Strategy on Integrated (Multi-Media) Pollution Control*, London: Institute for European Environmental Policy.

Jordan, A (2002) *The Europeanization of British Environmental Policy – A Departmental Perspective*, Basingstoke: Palgrave Macmillan.

OECD (1991) *Integrated Pollution Prevention and Control*, Environment Monographs No. 37, Paris: Organisation for Economic Co-operation and Development.

Owens, S (1990) The unified pollution inspectorate and best practicable environmental option in the United Kingdom. In: Haigh, N and Irwin, F, eds, *Integrated Pollution Control in Europe and North America*, Washington, DC: Conservation Foundation.

EU LEGISLATION MENTIONED IN THE TEXT

A *Directive* is binding as to the result to be achieved but leaves to the Member States the choice of form and method. A *Regulation* is directly applicable law. A *Decision* is binding in its entirety on those to whom it is addressed.

Directives

Regulations

ecolabel 880/92 71
eco-management and audit (EMAS) 1836/93 92, 172, 200
food safety authority 178/2002 157
neonicotinoid pesticides 485/93 159
ozone layer (CFCs) 3322/88 139
pollutant release and transfer register(E-PRTR) 166/2006 96
shipment of waste 259/93 69
vehicle emissions 715/2007 48, 56, 169, 194

Decisions

ALTENER (renewable energy) 93/500 100, 104–5
CFC's (ozone layer) 80/372 157–8
effort sharing of greenhouse gas reductions 406/2009 117
greenhouse gas monitoring mechanism 93/389, 280/200 28, 104, 106–108, 114, 118, 157
Kyoto Protocol ratification 2002/358 115
phthalates 1999/815 158
UNFCCC ratification of Convention 94/69 28, 101, 106

INTERNATIONAL CONVENTIONS AND THEIR PROTOCOLS MENTIONED IN THE TEXT

INDEX